T0299162

Bioreactor Modeling

Series Editor
Béatrice Biscans

Bioreactor Modeling

Interactions between Hydrodynamics and Biology

Jérôme Morchain

First published 2017 in Great Britain and the United States by ISTE Press Ltd and Elsevier Ltd

Apart from any fair dealing for the purposes of research or private study, or criticism or review, as permitted under the Copyright, Designs and Patents Act 1988, this publication may only be reproduced, stored or transmitted, in any form or by any means, with the prior permission in writing of the publishers, or in the case of reprographic reproduction in accordance with the terms and licenses issued by the CLA. Enquiries concerning reproduction outside these terms should be sent to the publishers at the undermentioned address:

ISTE Press Ltd
27-37 St George's Road
London SW19 4EU
UK

www.iste.co.uk

Elsevier Ltd
The Boulevard, Langford Lane
Kidlington, Oxford, OX5 1GB
UK

www.elsevier.com

Notices

Knowledge and best practice in this field are constantly changing. As new research and experience broaden our understanding, changes in research methods, professional practices, or medical treatment may become necessary.

Practitioners and researchers must always rely on their own experience and knowledge in evaluating and using any information, methods, compounds, or experiments described herein. In using such information or methods they should be mindful of their own safety and the safety of others, including parties for whom they have a professional responsibility.

To the fullest extent of the law, neither the Publisher nor the authors, contributors, or editors, assume any liability for any injury and/or damage to persons or property as a matter of products liability, negligence or otherwise, or from any use or operation of any methods, products, instructions, or ideas contained in the material herein.

For information on all our publications visit our website at http://store.elsevier.com/

© ISTE Press Ltd 2017
The rights of Jérôme Morchain to be identified as the author of this work have been asserted by him in accordance with the Copyright, Designs and Patents Act 1988.

British Library Cataloguing-in-Publication Data
A CIP record for this book is available from the British Library
Library of Congress Cataloging in Publication Data
A catalog record for this book is available from the Library of Congress
ISBN 978-1-78548-116-1

Printed and bound in the UK and US

Contents

Preface ix

Chapter 1. Tools for Bioreactor Modeling and Simulation 1

1.1. Introduction. 1
1.2. Process engineering approach 2
   1.2.1. Current modeling 2
   1.2.2. Multiphase modeling. 6
   1.2.3. Conclusion on the process
   engineering approach 18
1.3. Multiphase fluid mechanics approach 19
   1.3.1. General Euler equations 20
   1.3.2. Eulerian or Lagrangian approach
   for the biological phase? 22
   1.3.3. Conclusion on the computational
   fluid dynamics approach 26

Chapter 2. Mixing and Bioreactions 29

2.1. Introduction. 29
2.2. Mixing and reactions 31
   2.2.1. State of the problem 31
   2.2.2. Characterization of mixing 33
   2.2.3. Mixing mechanisms 44
   2.2.4. Characteristic mixing times 47
   2.2.5. Contribution of aeration to macromixing 48
   2.2.6. Reaction characteristic time. 52
2.3. Interaction between mixing and bioreaction 55
   2.3.1. Competition between mixing and
   chemical reaction. 55

2.3.2. Competition between mixing and
biological reaction 56
2.3.3. Spatial approach to the
micromixing problem 60
2.4. Analysis and modeling of couplings
between mixing and bioreaction 65
2.4.1. Link between the segregation
state and calculation of the apparent rate
of a simple biological reaction 66
2.4.2. Modeling of non-perfectly
mixed bioreactors. 71
2.4.3. Approach based on a mixing model 74
2.5. Conclusion 81

Chapter 3. Assimilation, Transfer, Equilibrium 85

3.1. Introduction. 85
3.2. Transfers between phases 86
3.2.1. Gas–liquid transfer 86
3.2.2. Liquid–microorganisms transfer 90
3.2.3. Synthesis and conclusion 96
3.3. Equilibrium or dynamic responses:
experimental illustrations 97
3.3.1. Dynamic response in terms of growth rate 98
3.3.2. Dynamic response for the assimilation 104
3.3.3. Dynamic response of the metabolism 108
3.4. Equilibrium models, dynamic models 113
3.4.1. Unstructured kinetic model: equilibrium
model or zero-equation model 114
3.4.2. Structured kinetic model. 116
3.4.3. Metabolic model 120
3.5. Confrontation of models with
experimental data 122
3.5.1. Response of a chemostat to an
increase in the dilution rate 123
3.6. Problem of coupling between a biological
model and hydrodynamic model 134
3.6.1. Closure of the liquid–cell transfer term 134
3.6.2. Problem of transport and mixing. 139
3.7. Conclusion 143

Chapter 4. Biological Population Balance 145

4.1. Introduction. 145

4.2. General population balance equation. 147
    4.2.1. Definitions. 147
    4.2.2. What are the links between structured
    models and population balances? 156
4.3. Illustrative examples 159
    4.3.1. Models based on mass discrimination. 159
    4.3.2. Differentiation by age 171
    4.3.3. Differentiation by intracellular composition 178
    4.3.4. Coupling with the bioreactor 179

Bibliography. 189

Index 205

Preface

The dynamic simulation of biological reactors is based on the modeling of a large number of coupled physical and biological phenomena. A predominant feature is the wide range of spatial and temporal scales involved. This modeling is based on the theoretical analysis of the phenomena and, when the techniques exist, on the analysis of experimental data. Research carried out for almost two decades now leads us to consider that turbulent flow, phase transfer, mixing state, biological reactions and the dynamics of microbial populations must be considered with the same interest and simultaneously with the extent of our means.

In the great majority of cases, biological reactors are agitated and/or aerated so that the flow regime is turbulent. However, in the field of reactive turbulent flows, particularly with regard to the respective contributions of fluid mechanics and process engineering, it is difficult to find a better introduction than the one given by Rodney Fox in his book *Computational Models for Turbulent Reacting Flows* [FOX 03].

Here are some excerpts:

At first glance, to the uninitiated, the subject of turbulent reacting flows would appear to be relatively simple. Indeed, the basic governing principles can be reduced to a statement of conservation of chemical species and energy... and a statement of conservation of fluid momentum... However, anyone who has attempted to master this subject will tell you that it is in fact quite complicated. On the one hand, in order to understand how the fluid flow affects the chemistry, one must have an excellent understanding of turbulent flows and of turbulent mixing. On the other hand, given its paramount importance in the determination of the types and quantities of chemical species formed, an equally good understanding of chemistry is required. Even a cursory review of the literature in any of these areas will quickly reveal the complexity of the task. Indeed, given the enormous research production in these areas during the twentieth century, it would be safe to conclude that no one could simultaneously master all aspects of turbulence, mixing, and chemistry.

Given their complexity and practical importance, it should be no surprise that different approaches for dealing with turbulent reacting flows have developed over the last 50 years. On the one hand, the chemical reaction engineering (CRE) approach came from the application of chemical kinetics to the study of chemical reactor design. In this approach, the details of the fluid flow are of interest only in as much as they affect the product yield and selectivity of the reactor. In many cases, this effect is of secondary importance, and thus in the CRE approach, greater attention has been paid to other factors that directly affect the chemistry. On the other hand, the fluid-mechanical (FM) approach developed as a natural extension of the statistical description of turbulent flows. In this approach, the emphasis has been primarily placed on how the fluid flow affects the rate of chemical reactions. In particular, this approach has been widely employed in the study of combustion...

In hindsight, the primary factor in determining which approach is most applicable to a particular reacting flow is the characteristic time scales of the chemical reactions relative to the turbulence time scales...

We recommend a thorough reading of this introduction and the first chapter which show how to establish a comprehensive dialogue between the two approaches. Naturally, our contribution will be much more modest but with an identical inspiration. Thus, in the first part of this work devoted to a synthesis of the various problems to be considered, we will use notions relating to process engineering, fluid mechanics, statistics and microbiology with the aim of highlighting convergence points. It will also be shown that each discipline meets its limits and the different views of each discipline make it possible to better understand the questions that are now receiving special attention in research.

We will then have the relationships of the characteristic times for the physical and chemical phenomena or biological phenomena; we will have approaches derived from fluid mechanics, reactor engineering and then from biology with all of its specificities. Among those, we can already emphasize that the particularity of bioprocesses (in relation to chemical processes) lies in what connects the living with the physical world, which are exchanges across the membrane.

In the introductory text of Rodney Fox, we can note the very close proximity between biological reactions and combustion. A fuel is a substance that, in the presence of oxygen and energy, can combine with oxygen (which acts as an oxidant) in a chemical reaction that generates heat: combustion. The parallel is obvious, even if it means using analogy, which is the weakest of logical arguments. The cell or microorganism, by virtue of its ability to produce energy, catalyzes the combustion of the carbonaceous substrate and oxygen which it collects from its environment. Feeding the cellular motor involves bringing fuel and oxidant to the cell and then passing them through the membrane, from the physical to the living. The two phenomena are necessarily consecutive: just as the mixture precedes the reaction, assimilation[1] is intercalated between these two in the case of microbiology. This interface, the membrane, which allows the internal biochemistry to be freed from the strict contingency of thermodynamics, constitutes a major interest for those studying the functioning of biological systems and their industrial applications. This singularity of the living being consisting of using the energy of intracellular reactions to control the transfer between the outside and the inside poses new difficulties compared to the case of chemical systems. While temperature, pressure and composition were sufficient to determine a rate of chemical reaction, it is necessary, in the case of bioreactions, to take into account the state of the cells. Everything would be only a little more complicated if the "state equations" of biological systems were known. However, the information at this level is often global averaged and we do not have local equations that are exact at the cell's scale, which can then be coupled with the flow and integrated to return to the scale of an elementary volume. On the contrary, the number of quantities required to determine the state of a microorganism is considerable. These difficulties

---

1 In other words, the transfer of matter between liquid phase and biological phase.

are at the heart of the modeling work for anyone who intends to couple hydrodynamics and biological reactions.

What else should be added to conclude this introduction? Beyond the order in which the various points (flows, transfers, bioreactions and mixing) will be approached, the last dimension relates to taking into account the notion of population, the diversity of states within the same population, the ability to remain outside the equilibrium with the external environment and the plasticity of the network of intracellular reactions. An important part will be devoted to the implementation of biological population balance and will propose, after a bibliographic synthesis, more forward-looking aspects.

It might seem obvious to use the concept of population balance in the study of biological systems and yet the number of works using the notion of population balance applied to bioprocesses remains low[2] (<1,000) whereas it exceeds 100,000 in the field of process engineering alone. There is therefore still much to be integrated, if not, to be invented in the field.

Jérôme MORCHAIN
July 2017

---

[2] Work in the field of medicine and mathematics should be included in order to be more realistic.

1

Tools for Bioreactor Modeling and Simulation

1.1. Introduction

The specificity of bioprocesses, in relation to chemical processes, lies essentially in the fact that reaction rates are not completely determined by thermodynamic variables (pressure, temperature, composition). Microorganisms constitute a phase in their own right and the apparent biological reaction rates are thus also a function of the microorganisms' state. However, this state is inherently a consequence of the environment in which the microorganisms evolve and that of the temporal variations in the concentrations seen by the cells along their trajectories. We find ourselves with bioreactors that have a two-way coupling: concentration fields and microorganisms influence each other. This specificity with respect to a conventional chemical system is shown in Figure 1.1.

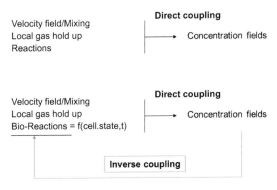

Figure 1.1. *Existence of inverse coupling is inherent in biological systems*

1.2. Process engineering approach

1.2.1. *Current modeling*

1.2.1.1. *General material balance equation*

Here, we resume the presentation as formulated by S.-O. Enfors in the book *Comprehensive Bioprocess Engineering* [BER 10] which is used as a reference work in the summer school of the European Federation of Biotechnology. In most bioprocess works, a formulation of material balances on a perfectly mixed reactor is found:

$$\frac{\partial V_m y}{\partial t} = F_i y_i + Q_i y_{G,i} - F_o y - Q_o y_{G,o} + r(y)V_m \qquad [1.1]$$

where:

– V_m is the volume of the culture medium;

– $r(y)$ is the reaction rate per unit volume (kg.m^{-3}.h^{-1}) for the production or consumption of the compound of concentration y;

– F_i and F_o are the flow rates of the culture medium at the inlet and the outlet;

– Q_i and Q_o are the gas flow rates at the inlet and the outlet.

Subscript G indicates that the compound belongs to the gas phase.

Under the assumption of a constant volume V_m, the general equation is simplified to give the overall mass balance that describes the evolution of concentrations over time:

$$\frac{\partial y}{\partial t} = \frac{F_i}{V_m} y_i - \frac{F_o}{V_m} y + \frac{\left(Q_i y_{G,i} - Q_o y_{G,o} \right)}{V_m} + r(y) \qquad [1.2]$$

The second term in the right-hand side, known as the gas transfer rate (GTR), represents the gas–liquid transfer rate per unit volume of the culture medium. In a closed reactor (batch): $F_i = F_o = 0$; in a semi-continuous reactor (Fed-batch): $F_i = F, F_o = 0$; and in a continuous reactor (chemostat): $F_i = F_o = F$.

1.2.1.2. *Partial material balance equations*

The generic variable y can be replaced by X, S, P or O_2 to designate the biomass (cells), the carbon substrate, a product of the biological reaction or the dissolved oxygen, respectively. Before establishing the balances for each compound, we need to provide an expression for the reaction rate and the transfer rate.

The *reaction rates* are expressed as:

$$r_j = q_j X \qquad\qquad [1.3]$$

where:

– X is the biomass concentration (kg.m^{-3});

– Q is the specific reaction rate (kg.kg$_{biomass}^{-1}$.h^{-1}), expressed per unit mass of cells;

– index j is used to identify the various compounds X, S, P or O_2.

The most commonly used form connects the specific rate q to the concentrations[1] and involves Y_{AB} yields in grams of A per gram of B.

$$q_X = \mu = \mu_{max} \frac{S}{K_s + S} \frac{O_2}{K_{O2} + O_2}$$
$$q_S = Y_{SX} q_X \qquad\qquad [1.4]$$
$$q_{O2} = Y_{O_2 X} q_X$$

The gas–liquid transfer rate is modeled in a conventional manner in the following form:

$$OTR = K_L a(O_2^* - O_2) \qquad\qquad [1.5]$$

The mass balance can then be written for the liquid phase for each of the compounds X, S and O_2 on a continuous reactor (chemostat):

---

1 q_X is generally denoted as μ: specific growth rate.

$$\frac{\partial X}{\partial t} = \frac{F}{V_m}\left(X_i - X\right) + q_{S,max}\frac{S}{K_s + S}X$$

$$\frac{\partial S}{\partial t} = \frac{F}{V_m}\left(S_i - S\right) - Y_{SX}\,\mu X \qquad\qquad\qquad [1.6]$$

$$\frac{\partial O_2}{\partial t} = \frac{F}{V_m}\left(O_{2i} - O_2\right) + K_L a\left(O_2^* - O_2\right) - Y_{O_2 X}\,\mu X$$

NOTE.– The formulation of q_X in the set of equations [1.4] indicates that the substrate and oxygen are both necessary for growth. An expression describing the growth rate $q_X = \mu(S_1) + \mu(S_2)$ is sometimes found if the described microorganism is capable of growing using two different substrates.

Linking the sugar and oxygen consumption rates algebraically to the growth rate presupposes a stabilized functioning of the microorganism population. This is known as *balanced growth*. As the term *balanced* refers here to the notion of equilibrium, it is essential to understand that there is an underlying notion of equilibrium in this common expression of material balance on a bioreactor.

1.2.1.3. *Critical analysis of classical modeling*

Several seemingly minor points deserve particular attention:

1) The balances are written on a liquid pseudo-phase representing the culture medium (with the cells).

2) The liquid pseudo-phase is assumed to be homogeneous (perfectly mixed).

3) The cells are seen as a dissolved species, which is not really the case because the culture medium is in fact a suspension. Adopting a dissolved species approach simplifies the problem formulation, but it ignores the underlying mass transfer between the liquid and the cells. Through this assumption, a shift from heterogeneous catalysis to homogeneous catalysis is carried out and the related issue of modeling the mass transfer between the cell and the liquid is put aside.

4) Volume V_m designates the volume of the medium. Therefore, the gas–liquid material volume transfer rate should be expressed per volume of medium. If the expression of this rate is calculated as the product of a coefficient K_L multiplied by an area of exchange a, attention must be paid

to the manner in which the latter quantity is calculated. In fact, the following equation is often used:

$$a = \frac{6\varepsilon_G}{d_{32}}$$ [1.7]

where a is the interfacial area per total contactor volume, i.e. by considering the total volume constituted from the culture medium and the bubbles. Within the framework of the modeling presented above, the calculation of the exchange area per volume of medium must therefore be done according to:

$$a = \frac{6\varepsilon_G(1-\varepsilon_G)}{d_{32}}$$ [1.8]

It will therefore also be necessary to devise a means of calculating the gas retention ε_G and the equivalent diameter of the bubbles d_{32}. This can be done using direct measurements, correlations or models.

5) The value of oxygen solubility, O_2^*, is assumed to be known. If the total quantity of oxygen lost by the gas phase is low, it can in practice be estimated that O_2^* is the value corresponding to equilibrium with the inlet gas phase $O_{2,Gi} = mO_2^*$. In a general case, it is necessary to add a material balance for oxygen in the gas phase and to solve it along with equation [1.6]. This means that a hydrodynamic model for the gas phase is required.

6) If an analogy is made with the combustion and operation of a petrol engine, the idea of a stabilized functioning of the microorganism population (or of equilibrium) may correspond to the notion of a functioning point of the engine with a particular fuel setting. The analogy stops there because the cell has the singular ability to adapt its fuel setting to the available resources. Thus, the set of equations [1.6] are valid for a population at equilibrium, i.e. having adopted the optimal fuel setting.

To conclude, the conventional model has the merit of simplicity, but it obscures the reality of certain physical phenomena and requires vigilance in the way in which certain quantities are calculated. At the same time, the comparison with the experimental results is not as direct as we might think. The approximations remain acceptable as they are of the same order as the experimental error, as long as the volume fractions of the gas and biological phases are small.

We should be cautious in the sense that treating the biological phase as a concentration in the material balance equations leads us to think of a biological reaction as a phenomenon related to homogeneous kinetics (between dissolved species) and to exclude aspects relating to transfers between the liquid and biological phases in the formulation of the problem. In reality, mixing, transfer and reaction intervene consecutively and the whole is indeed heterogeneous catalysis.

"In general, the rate equation for a heterogeneous reaction accounts for more than one process. This leads us to ask how such processes involving both physical transport and reaction steps can be incorporated into one overall rate expression. [...] In combining rates we normally do not know the concentrations of material at intermediate positions *(ndt: at the interface)*. Thus, we must express the rate in terms of the overall concentration difference" [LEV 72].

Finally, the cellular engine's efficiency is variable, which corresponds to different metabolic behaviors. This ability to modulate the intensity and nature of intracellular reactions gives microorganisms a great ability to adapt to environmental conditions. This implicitly raises questions about the nature of the driving force of this adaptation and the dynamics of the transition between two states of equilibrium.

1.2.2. *Multiphase modeling*

A more rigorous but also more complex approach consists of writing the global and partial mass balances at the reactor scale, in an integrated form, taking into account the triphasic character. The volume V designates the sum of the volumes of the three phases: liquid, gas and cells.

1.2.2.1. *Global material balance per phase*

$$\frac{\partial M_L}{\partial t} = \frac{\partial \rho_L V_L}{\partial t} = \rho_{L,i} F_{L,i} - \rho_{L,o} F_{L,o} + \Phi_{GL} + \Phi_{SL}$$

$$\frac{\partial M_G}{\partial t} = \frac{\partial \rho_G V_G}{\partial t} = \rho_{G,i} F_{G,i} - \rho_{G,o} F_{G,o} + \Phi_{LG} + \Phi_{SG} \qquad [1.9]$$

$$\frac{\partial M_S}{\partial t} = \frac{\partial \rho_S V_S}{\partial t} = \rho_{S,i} F_{S,i} - \rho_{S,o} F_{S,o} + \Phi_{LS} + \Phi_{GS}$$

where:

– ρ_k is the density of phase k (potentially a function of the composition);

– $F_{i,k}$ and $F_{o,k}$ represent the inlet or outlet volumetric flow rates for phase k;

– Φ_{kp} designates the net mass transfer rate from phase k to phase p.

Note that $\Phi_{kp} = -\Phi_{pk}$, so that the sum of the three equations of the system leads to the global mass balance on the reactor:

$$\frac{\partial M}{\partial t} = \frac{\partial\left(\varepsilon_L\rho_L + \varepsilon_G\rho_G + \varepsilon_S\rho_S\right)V}{\partial t} = \sum_{k=L,G,S}\left(\rho_{k,i}F_{k,i} - \rho_{k,o}F_{k,o}\right) \qquad [1.10]$$

1.2.2.2. Partial mass balances per phase

In the same way, a mass balance can be established for the compounds present in the different phases:

$$\frac{\partial \mathbf{m}_L}{\partial t} = \frac{\partial \varepsilon_L \mathbf{C}_L V}{\partial t} = F_{L,i}\mathbf{C}_{L,i} - F_{L,o}\mathbf{C}_{L,o} + \mathbf{R}_L\left(\mathbf{C}_L\right)V_L + \boldsymbol{\varphi}_{GL} + \boldsymbol{\varphi}_{SL}$$

$$\frac{\partial \mathbf{m}_G}{\partial t} = \frac{\partial \varepsilon_G \mathbf{C}_G V}{\partial t} = F_{G,i}\mathbf{C}_{G,i} - F_{G,o}\mathbf{C}_{G,o} + \mathbf{R}_G\left(\mathbf{C}_G\right)V_G + \boldsymbol{\varphi}_{LG} + \boldsymbol{\varphi}_{SG} \qquad [1.11]$$

$$\frac{\partial \mathbf{m}_S}{\partial t} = \frac{\partial \varepsilon_S \mathbf{C}_S V}{\partial t} = F_{S,i}\mathbf{C}_{S,i} - F_{S,o}\mathbf{C}_{S,o} + \mathbf{R}_S\left(\mathbf{C}_S\right)V_S + \boldsymbol{\varphi}_{LS} + \boldsymbol{\varphi}_{GS}$$

$$\varepsilon_L + \varepsilon_G + \varepsilon_S = 1$$

where:

– ε_k designates the mean volume fraction of the phase k;

– $\mathbf{C_k}$ designates the vector of mean volume concentration in phase k;

– $\mathbf{R}_k$ corresponds to the volumetric reaction terms in phase k. Each element of $\mathbf{R}$ corresponds to the net production rate of an element of C in phase k;

– $\boldsymbol{\varphi}_{kp}$ designates the vector mass fluxes from phase k to phase p per unit volume of phase p (kg.h^{-1}). Each element of $\boldsymbol{\varphi}$ corresponds to the transfer of an element of $\mathbf{C}$.

By taking the sum of the partial balances for all the compounds, by phase and then for all the phases, we will find the global mass balance.

However, the approach is limited for the biological phase because the exhaustive list and, to a lesser degree, the concentrations of all the intracellular constituents are not available.

NOTE.– No hypothesis on the homogeneity of the phases has been formulated so far. On the contrary, as the balance equations are written in integrated form, the terms of the reaction and transfer correspond to the sum of the local contributions on the volume of the phase in question. By introducing the notion of volume averaging $< >_k$ per phase:

$$R_k \equiv \langle R \rangle_k = \frac{1}{V_k} \iiint_{V_k} r(c_k) dv \qquad [1.12]$$

Modeling the reaction consists of finding a mathematical formulation of the term R_k giving an accurate approximate value of the volume integral of the reaction rates. In the simplest case, this term can be calculated using mean concentrations. Otherwise, a mixing model can be used to describe the concentration heterogeneity solved by the hydrodynamic[2] model and to explicitly calculate the reaction term. It should be noted that if the law $r(c)$ is not linear, it is wrong to model the reaction term in the form:

$$\langle R \rangle_k = r(\langle C \rangle_k) \qquad [1.13]$$

Concentrations follow the same logic and correspond to mean values per phase:

$$C_k \equiv \langle C \rangle_k = \frac{1}{V_k} \iiint_{V_k} c(v) dv \qquad [1.14]$$

The average volume fraction ε_k, defined in relation to the total volume, also results from a volume integral of the local retention α:

$$\varepsilon_k = \frac{1}{V} \iiint_V \alpha_k(v) dv \qquad [1.15]$$

---

2 We refer to a subgrid model in numerical fluid mechanics, but the notion remains valid in an ideal reactor.

The transfer terms correspond to the integral of the flux densities through the interfaces:

$$\phi_{kp} \equiv \iint\limits_{\Omega_{kp}} j_{kp} d\sigma \qquad [1.16]$$

The exchange surface Ω_{kp} depends on the volume fraction of the dispersed phases and the size distribution of the inclusions which need to be known. The explicit calculation of this term necessarily requires knowledge of the transfer law at the local scale. Otherwise, and this will be the case for the biological phase, a model based on the integrated quantities will be used. We will illustrate this particular point in the next section.

1.2.2.3. Implementation in the case of ideal reactors

The number of equations of the system formed by the total and partial balance equations per phase is less than the number of unknowns. Thus, it is necessary, in order to solve this problem, to give a certain number of relations that will allow us to close the system. The balances show the average volume fractions (or retentions) of the different phases. These values are generally considered to be constant for given operating conditions. Empirical correlations or considerations related to the slip velocities [COC 97] will be used to provide these values. In other words, we assume that the hydrodynamics obey a model of idealized flow in order to obtain a set of equations dealing only with concentrations. In addition, relations such as the relationship between density and composition $\rho_k = f(\mathbf{C}_k, P, T)$ should be added. Otherwise, the density must be considered to be constant and independent of the composition. We will now illustrate this approach for two particular configurations.

1.2.2.3.1. Perfectly mixed reactor (in all three phases)

In this ideal type of reactor, the volume fractions do not depend on the position within the reactor. Moreover, we assume that the liquid, gas and biological phases are perfectly homogeneous. Consequently, the composition of each phase is identical at all points of the reactor. Finally, we refer to the case of a reactor of constant total volume V.

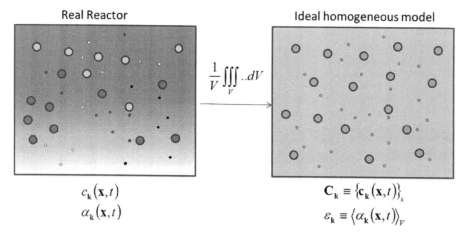

Real Reactor

$c_k(\mathbf{x},t)$

$\alpha_k(\mathbf{x},t)$

Ideal homogeneous model

$\mathbf{C_k} \equiv \{c_k(\mathbf{x},t)\}_k$

$\varepsilon_k \equiv \langle \alpha_k(\mathbf{x},t)\rangle_V$

Figure 1.2. *Schematic representation of the transition from a real configuration to a modeling approach based on the hypothesis of perfect mixing in the three phases. Here, the symbol **x** designates the position vector in space. For a color version of this figure, see www.iste.co.uk/morchain/bioreactor.zip*

The reaction and transfer terms can then be formulated from the concentrations in the different phases:

The reaction rates at the reactor scale are equal to the rates calculated from the average concentrations[3].

$$\mathbf{R}_k = \frac{1}{V_k}\iiint_{V_k} r(c)dv = r(\langle\mathbf{c}_k\rangle) \qquad [1.17]$$

Assuming a classical liquid–gas transfer model at the local scale, transfer rate multiplied by an exchange potential, we obtain:

$$\phi_{GL} = \iint_{\Omega_{GL}} K_L \Delta C d\sigma = K_L\left(C_L^* - C_L\right)\Omega_{GL} \qquad [1.18]$$

with $C_L^* = m.C_G$. We will note the interfacial area per reactor volume $a = \dfrac{\Omega_{GL}}{V}$.

---

3 This remains true in a volume that is not perfectly mixed if the reaction is of order 1.

The direct transfer from the gas phase to the biological phase is neglected $\varphi_{GS} = 0$.

The liquid–solid transfer term poses a real problem because, unlike in the gas–liquid case, there exists no "thermodynamic" or local equilibrium model to link the concentrations between the liquid and biological phases. The expression of the transfer can therefore only be made in a global sense, on the basis of the integrated or macroscopic quantities. For example, by carrying out a balance at the scale of an open, continuous, steady-state reactor, we have:

$$0 = F_L \mathbf{C}_{L,e} - F_L \mathbf{C}_L + \varphi_{SL}$$
$$\varphi_{SL} = -F_L \left(\mathbf{C}_{e,L} - \mathbf{C}_L \right)$$

[1.19]

This term will generally be expressed from a specific quantity **q** in grams (or moles) of the compound per gram of cell and per unit of time. However, this is indeed a *macroscopic* quantity whose calculation results from a balance at the scale of the reactor. The units of the specific quantity q suggest that it corresponds to a single cell property, that each cell realizes permanently, at the same rate as its neighbors, a part of the total mass transfer. It is quite practical but does not describe reality. Strictly speaking, **q** is an average value over the residence time of the reactor and over the entire population:

$$\mathbf{q} \equiv \overline{\langle q \rangle_S} = \frac{1}{T} \int \langle q \rangle_S \, dt$$

[1.20]

By dividing this flux by the existing mass of cells, the expected specific rate is produced. It will be noted that the ratio of the liquid flow rate to the volume of the reactor is homogeneous to the inverse of a time:

$$\mathbf{q} = \frac{\varphi_{SL}}{m_S} = -\frac{F_L \left(\mathbf{C}_{e,L} - \mathbf{C}_L \right)}{\rho_S \varepsilon_S V}$$

[1.21]

Rather than the volume fraction of the cells, the use has devoted the notion of cell concentration[4] (denoted as X in the medium g.l^{-1}) performing

---

4 Here, per volume of medium, or per volume of liquid, which slightly modifies the expression of q.

a shift from heterogeneous catalysis to the already mentioned homogeneous catalysis. Let us return to the definition of the cell concentration:

$$X = \frac{\rho_S \varepsilon_S V}{V_L + V_S} = \frac{\rho_S \varepsilon_S V}{(1 - \varepsilon_G) V_L / \varepsilon_L}$$

$$\mathbf{q} = -\frac{F_L (\mathbf{C}_{e,L} - \mathbf{C}_L)}{X (1 - \varepsilon_G) V_L / \varepsilon_L}$$

[1.22]

Note that $\mathbf{q}$ is a vector whose non-zero elements correspond to the compounds exchanged between the liquid phase and the biological phase. Moreover, even if it is the so-called "specific" quantity, it is not a local quantity.

It can be deduced that the material flow assimilated by the cells will most often be written in the form:

$$\phi_{SL} = \mathbf{q} X (1 - \varepsilon_G) V$$

[1.23]

The partial balances integrated on the scale of the reactor naturally make the terms appear similar to those of the conventional formulation but corrected for the volume fractions of different phases:

$$\frac{\partial \mathbf{C}_L}{\partial t} = \frac{F_{L,i} \mathbf{C}_{L,i} - F_{L,o} \mathbf{C}_L}{\varepsilon_L V} + r(\mathbf{C_L}) + K_{GL} \frac{a}{\varepsilon_L} (C_L^* - C_L) + \mathbf{q} \frac{1 - \varepsilon_G}{\varepsilon_L} X$$

$$\frac{\partial \mathbf{C}_G}{\partial t} = \frac{F_{G,i} \mathbf{C}_{G,i} - F_{G,o} \mathbf{C}_G}{\varepsilon_G V} + r(\mathbf{C_G}) - K_{GL} \frac{a}{\varepsilon_G} (C_L^* - C_L)$$

$$\frac{\partial \mathbf{C}_S}{\partial t} = \frac{F_{S,e} \mathbf{C}_{S,e} - F_{S,o} \mathbf{C}_S}{\varepsilon_S V} + r(\mathbf{C_S}) - \mathbf{q} \frac{1 - \varepsilon_G}{\varepsilon_S} X$$

$$\varepsilon_L + \varepsilon_G + \varepsilon_S = 1$$

[1.24]

1.2.2.3.2. Perfectly mixed reactor for the liquid and solid, plug flow for the gas

The configuration is shown in Figure 1.3. The only notable difference with respect to the preceding case is that it is necessary here to evaluate the integral of the oxygen flow transferred over the entire height H of the reactor and to express it from the data of the problem: F_G, C_{Gi}, H, $K_L a$ and C_L.

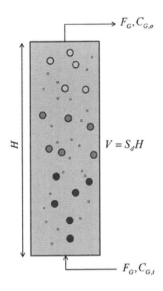

Figure 1.3. *Infinitely mixed bioreactor for the liquid and plug flow for the gas phase. Taking into account the depletion of the gas phase*

This time the flow of oxygen transferred depends on the altitude z, which varies between 0 and H. It is assumed that the gas flow is practically constant, which is reasonable if we consider, for example, that one transferred mole of oxygen will be compensated by one mole of CO_2 which will be desorbed from the liquid phase.

We can write, as in the previous case, an integral equation for oxygen, $O_{2,L}$ in the liquid phase at the reactor scale:

$$\varepsilon_L \frac{\partial O_{2,L}}{\partial t} = -q_{O_2} X + \frac{1}{V} \int_0^H K_L a \left(O_{2,L}^*(z) - O_{2,L} \right) S_d dz \qquad [1.25]$$

The difficulty thus consists of expressing the integral of the transfer flux from the variables of the problem: oxygen concentration in the liquid phase, flow rate and inlet and outlet concentrations of the gas phase. The trivial cases where the measurements of incoming and outgoing oxygen flow rates are available are omitted. The transferred flux is then calculated from the partial flow rates of oxygen at the inlet and the outlet. Rather than the value of the transferred flux, we seek to establish its mathematical expression here.

As the gas- and liquid-phase concentrations are linked by a thermodynamic equilibrium relation, it may be advantageous to establish the gas-phase concentration profile. We will see that it is not always necessary to solve this profile to express the overall oxygen mass transfer term.

From a balance on an elementary volume, of cross-section S_d and height dz, the variation of concentration in the gas phase in the absence of a reaction in the gas is written as:

$$\frac{\partial \varepsilon_G O_{2,G}(z) S_d dz}{\partial t} = F_G O_{2,G}\big|_z - F_G O_{2,G}\big|_{z+dz} - K_L a\left(O_{2,L}^*(z) - O_{2,L}\right) S_d dz \quad [1.26]$$

By integrating the height of the reactor, and assuming the total volume and the gas retention to be constant, this equation gives us:

$$\frac{\partial \langle O_{2,G} \rangle}{\partial t} = \frac{F_G O_{2,G,e} - F_G O_{2,G,o}}{\varepsilon_G V} - \frac{1}{\varepsilon_G H} \int_0^H K_L a\left(O_{2,L}^*(z) - O_{2,L}\right) dz \quad [1.27]$$

With the following definition of the average volume of the gas-phase concentration:

$$\langle O_{2,G} \rangle = \frac{1}{H} \int_0^H O_{2,G}(z) dz \quad [1.28]$$

There are two possible options for formulating the transfer term:

– the first consists of defining the mean value in the gas phase as the arithmetic mean of the input and output concentrations:

$$\langle O_{2,G} \rangle = \frac{O_{2,G,e} + O_{2,G,s}}{2} \quad [1.29]$$

This assumption is reasonable when the flux of transferred oxygen is small compared to the total injected flux.

It will be observed that the transfer term is a linear function of the equilibrium O_2^* concentration. Therefore, if the concentration in the gas varies linearly with the altitude z, the expression proposed above will be exact and the integral of the transfer flux corresponds exactly to the flux calculated from the average equilibrium concentration $\langle O_{2,L}^* \rangle = m \langle O_{2,G} \rangle$.

By injecting this last relation and the proposition of equation [1.29] into equations [1.25] and [1.27], we obtain a closed system of two equations for the outlet concentration in the gas phase $\langle O_{2,G,s} \rangle$ and the concentration in the liquid phase $\langle O_{2,L} \rangle$:

$$\frac{\partial O_{2G,s}}{\partial t} = \frac{F_G O_{2,G,e} - F_G O_{2,G,s}}{\varepsilon_G V} - \frac{K_L a}{\varepsilon_G}\left(m\frac{O_{2,G,e} + O_{2,G,s}}{2} - O_{2,L} \right)$$

$$\frac{\partial O_{2,L}}{\partial t} = -\frac{q_{O_2} X}{\varepsilon_L} + \frac{K_L a}{\varepsilon_L}\left(m\frac{O_{2,G,e} + O_{2,G,s}}{2} - O_{2,L} \right)$$

[1.30]

EXAMPLE.– Suppose that the oxygen concentration in the gas phase changes linearly with the altitude in the column:

$$O_{2,G}(z) = \alpha.z + \beta$$

[1.31]

Let us express the average volume concentration of the gas phase as:

$$\langle O_{2,G} \rangle = \frac{1}{H}\int_0^H (\alpha.z + \beta)dz$$

[1.32]

We thus obtain:

$$\langle O_{2,G} \rangle = \frac{1}{H}\int_0^H (\alpha.z + \beta)dz = \frac{1}{H}\left[\frac{\alpha.z^2}{2} + \beta z + \delta \right]_0^H = \alpha\frac{H}{2} + \beta = O_{2,G}\left(z = \frac{H}{2} \right)$$

[1.33]

Thus, the average concentration volume of the gas phase corresponds to the mid-height value.

On the contrary, if we express the half-sum of the gas inlet and outlet concentrations, we obtain:

$$\frac{O_{2,G,e} + O_{2,G,s}}{2} = \frac{\alpha H + \beta + \alpha.0 + \beta}{2} = \alpha\frac{H}{2} + \beta$$

[1.34]

CONCLUSION 1.– If the oxygen concentration in the gas phase is a linear function of the height, the average concentration in the gas phase is identical to the half-sum of the inlet and outlet concentrations.

We now prove that the integral of the transfer term in equation [1.27] can be written from the average volume concentration $\langle O_{2,G} \rangle$ that has just been calculated:

$$\int_0^H K_L a \left(O_{2,L}^*(z) - O_{2,L} \right) dz = K_L a \int_0^H \left(m(\alpha.z + \beta) - O_{2,L} \right) dz \qquad [1.35]$$

which leads by integration to:

$$K_L a \left[m\alpha \frac{z^2}{2} + \left(m\beta - O_{2,L} \right) z + c \right]_0^H = K_L a \left(m\alpha \frac{H^2}{2} + \left(m\beta - O_{2,L} \right) H \right) \qquad [1.36]$$

Moreover, given that $<O_{2,G}>$ and $O_{2,L}$ are inherently independent of the altitude z, we have:

$$\int_0^H K_L a \left(m\langle O_{2,G} \rangle - O_{2,L} \right) dz = K_L a \left(m\langle O_{2,G} \rangle - O_{2,L} \right) H \qquad [1.37]$$

We thus have equality between the left-hand side term of equation [1.35] and the right-hand side term of equation [1.36], justifying the transition from equation [1.27] to equation [1.30] that was previously realized.

CONCLUSION 2.– The integral of the transferred oxygen flux can be expressed from the average concentration in the gas phase.

– The second option consists of establishing the exact law of variation of the concentration in the gas phase as a function of altitude. Subsequently, this expression will be used to calculate the integral of the oxygen flux transferred to the liquid phase and thus close the system of equations. Equation [1.38] presents the mass balance for oxygen in the gas phase over an elementary volume of height dz at the steady state:

$$\frac{\partial O_{2,G}(z)}{\partial z} = -\frac{K_L a}{F_G / S_d} \left(m O_{2,G}(z) - O_{2,L} \right) \qquad [1.38]$$

This leads, after integration, to the expression of the equilibrium concentration in the liquid phase as a function of the altitude z:

$$\frac{mO_{2,G}(z) - O_{2,L}}{mO_{2,G,e} - O_{2,L}} = e^{-m\frac{K_L a}{F_G/S_d}z} \qquad [1.39]$$

It can be seen here that the hypothesis of a linear variation of the concentration with relation to the altitude constitutes a very useful approximation to simplify the writing of the transfer term but proves to be inaccurate when the ratio between the residence time of the bubbles $S_d H/F_G$ and the characteristic time transfer, $1/K_L a$, is large. In other words, the linear approximation is only valid if the quantity transferred to the liquid is small. By injecting expression [1.39] into the integral term in equation [1.25], we obtain the following equation:

$$\int_0^H K_L a\left(O_{2,L}^*(z) - O_{2,L}\right) dz = \int_0^H K_L a\left(mO_{2,G,e} - O_{2,L}\right) e^{-m\frac{K_L a}{F_G/S_d}z} dz \qquad [1.40]$$

This expression is further integrated to obtain the total amount of oxygen transferred by unit time and unit volume of the reactor:

$$\frac{1}{H}\int_0^H K_L a\left(O_{2,L}^*(z) - O_{2,L}\right) dz = \frac{F_G}{mV}\left(O_{2,L,e}^* - O_{2,L}\right)\left[1 - e^{-m\frac{K_L a V}{F_G}}\right] \qquad [1.41]$$

Now, let us replace the term representing the transfer in equation [1.25] with the expression that has just been established to obtain a single equation describing the evolution of the oxygen concentration in the liquid phase:

$$\varepsilon_L \frac{\partial O_{2,L}}{\partial t} = -q_{O_2} X + \frac{F_G}{mV}\left(O_{2,L,e}^* - O_{2,L}\right)\left[1 - e^{-m\frac{K_L a V}{F_G}}\right] \qquad [1.42]$$

It is easy to show that the right-hand side of equation [1.41] can be rewritten as a macroscopic balance of the gas phase oxygen. The measurement of the flows and concentrations on the gas phase at the inlet and the outlet makes it possible to obtain this value. However, the expression for the total mass transfer [1.41] is obtained by integrating a local transfer law; it is therefore predictive because it is expressed from the model's input variables (incoming gas flow rate, concentrations, transfer coefficient, equilibrium constant and reactor volume).

CONCLUSION.– We thus conclude with these two examples that the knowledge of the law of the liquid–gas transfer on the local scale allows for the explicit calculation of the integral of the transfer flux in a predictive way. There is therefore a fundamental difference between the modeling of the gas–liquid transfer term and that of the liquid–cell transfer for which it is not possible to express the local transfer law.

1.2.3. Conclusion on the process engineering approach

The traditional approach in process engineering/reactor engineering addresses process modeling at the scale of the reactor or as a set of interconnected zones (compartment model approach). If the compartment model approach is chosen, mass fluxes as well as volume fractions in each zone should be given through correlations or experimental data [VRÁ 00]. In fact, local phenomena are not explicitly described and must be modeled. Among these phenomena, the case of gas–liquid transfer is well known and the production of an integrated model is facilitated by the knowledge of local physics at the interface. This is not the case for the biological phase, as we will see later.

The approach inspired by process engineering (or reactor engineering) finds its limit when internal hydrodynamics and transfer phenomena or reactions are strongly coupled. It then becomes very difficult to produce an integrated model that accounts for the multiple interactions occurring on a small scale compared to that of the reactor. This is precisely what happens in the case of bioreactors: global functioning is the result of all the fluctuations experienced by the biological systems along their trajectories. All this happens on microscopic scales (that of microorganisms) but this time we do not have the physical laws applicable at this scale. It is therefore practically impossible to integrate these laws to produce a model at the reactor scale. As a result, a number of empirical laws (we could say correlations) that connect operating variables (flow rates, concentrations) to the overall apparent rates of biological reactions have been produced from the experimental results. Monod's law is a good example of this: at the steady state and on average over the population, the growth rate (average of the population) is equal to the dilution rate. In addition, there is a relationship between the growth rate and the residual concentration of the limiting substrate. This is a macroscopic observation (at the reactor scale). However, it is rather doubtful to think that all the organisms grow at exactly the same rate. Thus, this law, while valid macroscopically, cannot be inferred at the scale of an individual. Furthermore, it is not valid in dynamic regime when the variations imposed on the reactor are too sudden. The predictive capacity of these integrated correlations is therefore limited.

How can we go beyond these? More local formulations should be sought for flows, transfers and reactions.

1.3. Multiphase fluid mechanics approach

The use of the multiphase fluid mechanics approach becomes essential when it is necessary to explicitly calculate the reactor's internal hydrodynamics, concentration fields, and transport of the microorganisms thus approaching the cell scale for more accurate modeling of local phenomena. The solution of these equations then calls for numerical fluid mechanics. We will examine, later in this book, the conditions that make such an approach necessary. Let us note already that the main question is that of the transport of the biological phase and the means to track the events encountered by the microorganisms. Let us note that this question arises as soon as we are interested in a cascade of reactors or a recycle reactor. Figure 1.4 shows the different levels of hydrodynamic modeling and cites some representative works in the field of numerical simulation of bioreactors. We emphasize on the fact that the increase in the spatial resolution (splitting of the reaction volume into smaller and smaller sub-volumes) is accompanied by an increase in the temporal resolution. The smaller the spatial scale is, the smaller the residence time in a volume will be. This refinement, this precision in space, only has meaning if we can describe the transfer and the biological reactions on the same scale. Of what use would it be to know the history of the concentrations seen by a cell at the millisecond scale if the biological model responds at the minute scale? At a minimum, it would be appropriate, in such a situation, to integrate the fluctuations over one minute before invoking the biological model. What impact can the instantaneous variations with respect to this average have on the cell's functioning?

Reactor Engineering   Computational Fluid Dynamics

Ideal Reactor   Reynolds Averaged Navier-Stokes equations (RANS)
Compartment Model   Large Eddy Simulation (LES)
+ micromixing model   Direct Numerical Simulation (DNS)

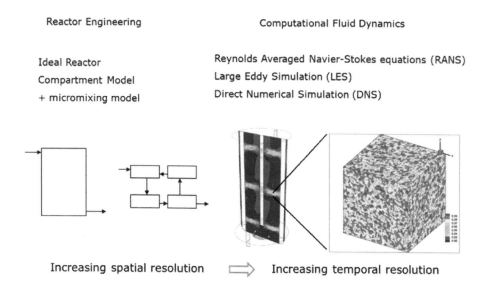

Increasing spatial resolution   ⟹   Increasing temporal resolution

Figure 1.4. *Various hydrodynamic modeling tools for bioreactors. For a color version of this figure, see www.iste.co.uk/morchain/bioreactor.zip*

The works integrating the reactive aspects are listed below; those in which this aspect is absent are marked with a star.

– Compartment models: Mayr *et al.* [MAY 94, MAY 93], Vràbel *et al.* [VRÁ 01], Delvigne *et al.* [DEL 08b, DEL 05], Morchain [MOR 00] and Pigou and Morchain [PIG 15].

– Reynolds-averaged Navier–Stokes equations (using a turbulence model): Hjertager [HJE 95], Bezzo *et al.* [BEZ 05, BEZ 03], Hristov *et al.* [HRI 01, HRI 04], Lapin *et al.* [LAP 06, LAP 04], Morchain *et al.* [MOR 13a, MOR 12], Schmalzriedt *et al.* [SCH 03] and Schütze and Hengstler [SCH 06]. *Zhang *et al.* [ZHA 09] and *Moilanen *et al.* [MOI 07, MOI 05].

– Direct numerical simulation: Linkès *et al.* [LIN 12b, LIN 14].

1.3.1. *General Euler equations*

The equations for the conservation of the total mass, the momentum per phase and the mass per phase on an elementary control volume are given by:

$$\frac{\partial \alpha_k \rho_k}{\partial t} + \nabla \alpha_k \rho_k \mathbf{U}_k = 0$$

$$\frac{\partial \alpha_k \rho_k \mathbf{U}_k}{\partial t} + \nabla \alpha_k \rho_k \mathbf{U}_k \mathbf{U}_k = -\nabla \alpha_k P_k + \nabla \alpha_k \left(\overline{\tau_k} - \rho_k \overline{u'_k u'_k} \right) + \overline{M_k} + \alpha_k \rho_k \mathbf{g} \quad [1.43]$$

$$\frac{\partial \alpha_k c_k}{\partial t} + \nabla \alpha_k c_k \mathbf{U}_k = \alpha_k S_k - \nabla \alpha_k \left(\overline{J_k} + \overline{c'_k u'_k} \right) + \overline{L_k}$$

where:

– ρ_k: density of phase k;

– α_k: local volume fraction of phase k;

– U_k: velocity vector of phase k;

– c_k: concentration of compound c in phase k. There are as many equations of this type as there are compounds per phase;

– M: interfacial momentum exchange;

– L: interfacial diffusive exchange of species c.

In the particular case of bioreactors, index k designates the liquid, gaseous or biological phase. Although it can be considered as the most natural one, there is no trace of a general formulation using this Euler tri-phase approach in the literature dealing with bioreactors.

It is certainly not always necessary to refer to such a complete formulation to tackle the modeling of bioreactors. A simplified formulation as it is implemented in the field of reactor engineering may be suitable for many practical situations. However, the fluid mechanics-based approach has the merit of explicitly revealing the different terms related to distinct physical phenomena and invites reflection on the associated characteristic scales. Thus, a certain number of questions arise, particularly at the level of the exchange terms between phases:

Is there a possibility of direct mass transfer from the gas phase to the biological phase?

Is the transfer between the gas and the liquid affected by the presence of the microorganisms, in a purely physical manner or according to a process analogous to acceleration by the reaction?

Is there a limitation of the reaction in the biological phase by the transfer; should we then distinguish (as in heterogeneous catalysis) a physical regime and a biological one?

The idea of thermodynamic equilibrium at the interface is generally accepted for exchanges between a gas and a liquid. Moreover, this equilibrium is instantaneous. Can we admit in the same way the idea of an instantaneous equilibrium between the composition of the liquid phase and the biological phase? If not, how long does it take to reach equilibrium?

1.3.2. *Eulerian or Lagrangian approach for the biological phase?*

Concerning the gas–liquid aspect, the Eulerian approach for two fluids is used. It remains to be seen how to take the biological phase into account. In most bioreactors, the biological phase is suspended in the liquid phase in the form of isolated cells, biological aggregates or biofilm developed on carrier solids.

To process the biological phase, two approaches can be envisaged:

– Lagrangian approach: the cells are treated as "solid" particles suspended in the liquid. The individual trajectories are calculated from a balance of forces on the particle. This approach was implemented for bioreactors by Matthias Reuss *et al.* [LAP 06, LAP 04]. The mathematical expression for each force (weight, drag, lift, added mass) depends on the scale chosen for the resolution of the flow of the continuous phase.

– Eulerian approach: the cells are taken into account through their volume fraction in the mixture. There is no distinction between the different individuals present in an elementary volume. This model gives access to the volume fractions and the velocities of each phase at any point of the reactor. The transfer and reaction terms must be modeled at the level of the elementary control volume.

What are the advantages and disadvantages of these approaches?

On this basis, the question deserves, above all, to be posed beyond the hydrodynamic aspect. We will, however, begin by examining this point. For microbial cells in the µm scale whose densities are close to that of water, the calculation of the Stokes number shows that the particles behave like tracers of the fluid particles [LIN 14]. It is therefore reasonable to neglect the term of momentum exchange between the cells and the fluid as the slip velocity between the phases is zero.

Thus, *for momentum*, the equation of the three-phase Eulerian model relative to the solid phase can be simplified and the microorganisms can be represented by a single variable, their volume fraction, and transported by the liquid phase.

The Stokes regime also leads to important simplifications of the cell trajectory equation. However, the Lagrangian approach poses two problems, one of the problems is of a practical nature, linked to the management of voluminous files containing the particle trajectories; the other related to the modeling of the effect of turbulence on the particle trajectory. If the Reynolds-averaged approach is used to predict the velocities of the continuous phase, a correct trajectory calculation should involve a model to calculate the turbulence-induced velocity fluctuations (this problem disappears in the case of a DNS simulation, see M. Linkès' thesis [LIN 12b]) From the hydrodynamic point of view, the Eulerian approach is ultimately the simplest to implement.

Let us examine the question from the point of view of the transfer and reaction aspects' modeling. Clearly, the Lagrangian approach offers wider possibilities in this area. It gives access, by nature, to individual information. This makes it possible to follow the position and composition of each cell, to realize ensemble averages and to produce statistical information at the population scale (probability of finding an individual with a given composition within the population of the reactor). Given the number of suspended microorganisms in a bioreactor, it is not possible to carry out an individual monitoring of each microorganism. A volume fraction of 0.1%, approximately 1 g.l^{-1}, corresponds to 10^{12} cells.cm^{-3}. In practice, the monitoring of particles (representing a large number of real cells) is thus carried out and the average behaviour described using a cell-based model. The main interest of the Lagrangian approach is to allow for the use of a very large number of variables to describe the functioning of the biological system. It is thus very easy to inject, in the flow's numerical simulation, the biological model established on the scale of a laboratory bioreactor, regardless of the complexity of the said model because the necessary variables are attached to the particle. The effect of concentration variations seen along the trajectory on the biological system (known as direct coupling) is fairly easily accessible by the Lagrangian approach.

The treatment of the reactive aspect in an Eulerian approach poses *a priori* another type of problem. Let us begin with what will be similar to a Lagrangian vision: we can attach a certain number of variables to the dispersed phase (here the cells) allowing us to describe the biological functioning of the microorganism. A simple (kinetic) model or a more refined model (metabolic, or metabolic/cybernetic) can be envisaged. The difficulty that appears in the Eulerian approach relates to the question of mixing. Let us imagine two containers each containing cells in different states. The variables attached to the biological phase initially have different values in each container. In practice, when mixing these two volumes, each group of cells would keep its own state (at least at the beginning before it can evolve). In no case, the contents of the cells from two containers would be mixed to obtain an average value. This is what happens when using a multiphase Eulerian approach: the content of each phase is homogenized. This may be relevant for bubbles or drops because the rupture/coalescence phenomena do indeed homogenize the contents but this is not the case for solid particles that do not fuse together.

A difficulty arises with the Lagrangian approach when we want to describe the modification of the liquid phase's concentration field due to the assimilation or excretion by the microorganisms (known as inverse coupling). With the use of numerical fluid mechanics tools, the reactor's volume was divided into a large number of control sub-volumes in order to calculate the velocities at all the points. In the Eulerian approach, the gas and biological phase fractions are also accessible in each of the control volumes. With a Lagrangian approach, it is necessary to ensure the presence of at least one particle per control volume in order to be able to estimate the production/consumption term in the liquid phase. In the case of an agitated and aerated industrial bioreactor, we have here a few million Lagrangian particles. The flow/transfer/reaction coupling in the case where the biological solid phase is constituted of a heterogeneous population necessitates a sufficient number of distinct individuals in each control volume to ensure the statistical convergence of the ensemble average of the particles. This condition, necessary to precisely calculate the reaction term, leads to a roadblock for the Lagrangian approach to be applied to the case of large bioreactors. On the contrary, when applied to smaller volumes (scale of a control volume for example), it can be very useful for carrying out numerical experiments from which models to be injected into the Eulerian approach are built.

	Eulerian	Lagrangian
Direct coupling	Simple	Simple
Inverse coupling	+ Phase transfer terms are explicit	− Requires a high number of particles per control volume (mesh)
Biological model	+ Equilibrium model − Structured models	+ No limitation in complexity or number of variables
	The temporal resolution of the biological model must be adapted to the simulated concentration fluctuations	
Population heterogeneity	+ Accessible via the addition of a population balance equation − Limited number of variables	Related to the number of particles + The global heterogeneity (at the reactor scale) accessible starting from few thousands of particles
Numerical	Stationary or non-stationary calculations Specific methods for BP	High post-treatment cost Non-stationary calculations
Historical	No direct access to the particles' history	Individual knowledge of trajectories. Frequency analysis possible

Table 1.1. *Comparative advantages of the Eulerian and Lagrangian approaches for the numerical simulation of a bioreactor*

In the Eulerian approach, the biological phase is represented by the volume presence rate (or concentration) and a concentration vector (in the biological phase). It should then be kept in mind[5] that an ensemble average of all individuals has been taken. The representation by the volume fraction means that the behavior of the population is described through an average individual of average composition. There are many implications to this especially because biological systems are nonlinear, the dynamics of a population is not that of the average individual. In order to regain the population scale, it is appropriate to add to the Eulerian approach a population balance approach for the biological phase. By its very nature, the addition of a population balance equation makes it possible to manage the question of mixing mentioned above. This approach is, in our view, superior to the Lagrangian approach in the case of industrial reactors.

---

5 In modeling as well as for the analysis of experimental results.

The basic problem thus concerns the writing of a model based on the Eulerian variables which integrates the various underlying physical and biological phenomena. The question of the liquid–cell mass transfer under the resolved scale (linked to the choice of the hydrodynamic model) therefore remains unresolved. Moreover, this writing requires usage of the multi-variable population balance formalism to allow for the association with a structured biological model.

1.3.3. *Conclusion on the computational fluid dynamics approach*

The multiphase equations formulated in an Eulerian way are derived from volume averaging and therefore cover velocities and volume fractions per phase. The solution of these equations requires the use of numerical fluid mechanics calculations making it possible to approach the study of industrial reactors. In this simulation step, the volume of the reactor is divided into a large number of elementary volumes (meshes or control volumes) for which the basic equations are integrated. The local equations are therefore filtered spatially and the phenomena whose characteristic scale is smaller than the size of the mesh must then be modeled using the law of closure known as *subgrid scale models*.

The first example that comes to mind is that of the modeling of velocity fluctuations due to the turbulent characteristics of the flow; various turbulence models are available in most CFD codes. We can also think immediately of the existence of a concentration distribution within an elementary volume and the consequences on the expression of a biological reaction term integrated on the scale of a computational mesh. Finally, it should also be noted that the population of microorganisms within a computational mesh is not necessarily homogeneous, which also has consequences for the calculation of the reaction and/or phase exchange term.

Thus, it is important to note that the filtering used for hydrodynamics (Reynolds-averaging, phase average, volume average) is required for the other terms of the conservation equation. Subgrid scale models for different terms can be proposed. For example: interaction by exchange with the mean (IEM) model for mixing and population balance model for the biological phase.

To conclude, a three-phase Eulerian approach is the appropriate framework for a number of theoretical reflections. It is also adapted to the integration of models of the population balance type used to describe the heterogeneities of the gas and biological phases. This involves modeling physical and biological phenomena based on the knowledge of local mechanisms. These steps are necessary to obtain a sufficiently complete description of the couplings within an industrial size bioreactor. With the Lagrangian approach, the integration of biological models is more natural, but the difficulty relates to the expression of the transfer and reaction terms. A central question arises in both cases: by what mechanisms and with what dynamics, does the biological system respond to external fluctuations? Part of the answer to this question lies in the transfer between the liquid and biological phases.

2

Mixing and Bioreactions

2.1. Introduction

The problem related to the state of mixing lies in the calculation of the source/sink term due to the biological reaction. Whatever the approach used to model the flow of the different phases and the transport of the dissolved species, there comes a time when it becomes necessary to calculate a term that integrates all the exchanges between the culture medium and the biological phase. The formulation of this term depends on whether heterogeneities are taken into account within each phase. Thus, the three stages of substrate assimilation, their use within microbial catalysts (which can lead to the multiplication of individuals) and the excretion of products are most often grouped under a single term called "biological reaction". The underlying mechanisms are the mass transfer between phases and a set of interdependent reactions within the biological phase. In this chapter, however, we will use an integrated view in which the biological phase is treated as a species within the solution, allowing us to use analogies with studies on the interactions between mixing and chemical reactions in a homogeneous phase. Let us keep in mind that this is a simplification to which we must return.

The amount of experimental evidence showing the effect of mixing on biological reactions is significant [AL 07, AMA 01, BAJ 82, BYL 98, DEL 15, DUN 90, HAN 66, LAR 06, LIN 00]. It is now understood that poor mixing reduces biomass yields and promotes the production of by-products. It is also clear that living microorganisms respond to fluctuations in concentrations along their trajectory in the bioreactor and that the observed behavior is an integrated consequence of these fluctuations. It is also a consequence averaged over a very large set of individuals.

Analysis of the interactions between mixing and bioreactions remains complex for several reasons:

– the flow regime in bioreactors is generally turbulent. One thus touches on a field that is complex in itself and falls under fluid mechanics;

– the micrometric size of industrially grown cells is comparable to the smallest scales of turbulence;

– the range of biological phenomena by which a microorganism can respond to fluctuations in concentrations is very wide. It is therefore difficult to define with certainty the rate of the reaction impacted by the mixing;

– ticroorganisms adapt to their environment so that the very nature of the response can change over time. Short- and long-term responses can be distinguished;

– the biological reactor is not a homogeneous system and the direct transposition of the concepts of chemical reactor engineering to the case of a bioreactor proves to be difficult.

The approach we propose here consists of studying the mixing from a purely physical point of view and then providing the basic tools for studying the link between the apparent rate of a chemical reaction and the state of mixing in the case of homogeneous chemical systems. We recommend reading Mann et al. [MAN 95] as an introduction. Finally, we shall conclude with the case of biological reactions.

This chapter will illustrate the necessary coherence of the representative models of hydrodynamics and bioreactions. The study of the interactions between mixing and biological reactions can be considered at different scales, from that of the cell to that of the entire reactor. At these spatial scales, there are corresponding temporal scales. It is therefore very useful to grasp some fundamental notions that determine the history of the cellular environment at all these scales. We will illustrate this from examples from the literature and personal contributions.

2.2. Mixing and reactions

2.2.1. *State of the problem*

The objective is to obtain a formulation of the reaction term on a macroscopic scale while integrating the effect of the interactions that occur on smaller scales. When the mass balance is established on a reagent in a reactor of volume V, it should be borne in mind that:

– In the accumulation term, the evolution of the total mass of compound C is expressed as the product of the volume by a concentration that corresponds in fact to a volumetric average <C>.

$$\langle C \rangle = \frac{1}{V} \iiint_V C(v)dv \qquad\qquad [2.1]$$

– In the reaction term, the volume integral of the local reaction rates r is expressed.

We should therefore write the conservation equation in the following form:

$$\frac{\partial \langle C \rangle V}{\partial t} = \iiint_V r_C(\mathbf{C}(v))dv \qquad\qquad [2.2]$$

where C(v) is the local concentration vector in a homogeneous elementary volume *dv*.

By defining <r_C> as the volume average of the reaction rate:

$$\langle r_c \rangle = \frac{1}{V} \iiint_V r_C(\mathbf{C}(v))dv \qquad\qquad [2.3]$$

We get an exact equation

$$\frac{\partial \langle C \rangle V}{\partial t} = \langle r_c \rangle V \qquad\qquad [2.4]$$

Since this equation describes the evolution of the average concentration, the reaction term must be expressed from the average concentration as well. Considering that the local kinetic law is known, the simplest solution consists of stating that:

$$\left\langle r_C \right\rangle = r_C \left(\left\langle \mathbf{C} \right\rangle \right) \tag{2.5}$$

It is now necessary to identify the conditions in which this modeling is accurate and, conversely, those in which an error is made by adopting it. For this purpose, we will consider that the volume V is not homogeneous but that it can be subdivided into a large number of elementary volumes that are homogeneous. The concentration in an elementary volume can be written as the sum of the average value and a fluctuation (deviation from the mean):

$$\mathbf{C}(v) = \left\langle \mathbf{C} \right\rangle + \mathbf{c}(v) \tag{2.6}$$

Taking the volume average of the expression [2.6] leads directly to the fact that the average of the concentration fluctuations is zero, $<c(v)>=0$.

Let us now assume a kinetic law of order 1, $r(C)=kC$. Let us calculate the apparent reaction rate at the scale of the entire volume. We get:

$$\left\langle r(C) \right\rangle = \left\langle k\left(\left\langle C \right\rangle + c(v)\right) \right\rangle = k\left\langle C \right\rangle \tag{2.7}$$

The average reaction rate is equal to the reaction rate calculated from the average concentration. Even if concentration heterogeneity exists in the reactor, we just need to know the average value of the concentration to accurately calculate the resulting overall rate at the reactor scale.

On the other hand, if the reaction is of order 2, $r(C)=kC^2$, we obtain:

$$\left\langle r(C) \right\rangle = \left\langle k\left(\left\langle C \right\rangle^2 + 2\left\langle C \right\rangle c(v) + c(v)^2\right) \right\rangle ...$$
$$\left\langle r(C) \right\rangle = k\left\langle C \right\rangle^2 + k\left\langle c(v)^2 \right\rangle = r\left(\left\langle C \right\rangle\right) + k\left\langle c(v)^2 \right\rangle \tag{2.8}$$

Since the average of the squared fluctuations is not zero, the mean velocity is not equal to the velocity calculated from the average concentration. When using equation [2.5] in order to model the reaction term, the more the concentration field is heterogeneous, the greater the calculation error.

CONCLUSION.– The apparent reaction rate is affected by the state of mixing when the kinetics are of an order greater than 1. The biological kinetics, even when described using the simplest model (Monod's law, for example), are not linear. They are therefore sensitive to the mixing state: the average reaction rate is not equal to the velocity based on the average concentration.

NOTE.– It would be wrong to consider this conclusion only from the modeling point of view. Indeed, to analyze the experimental results by making the hypothesis of a perfect mixing, whereas the reactor is in practice heterogeneous, will lead to erroneous calculation of the kinetic constant. It is easy to show that, in this situation, the real reaction constant k and the apparent constant k_{app} (associated with the perfect mixing hypothesis) are related by equation $k_{app} = k \left(1 + \frac{\langle c(v)^2 \rangle}{\langle c \rangle^2} \right)$.

NOTE.– The simplification of equation [2.4] as $\frac{\partial c}{\partial t} = r_c$ is a source of many errors and other difficulties of interpretation. We urge the reader not to proceed with this simplification without having studied the consequences with the utmost care.

2.2.2. Characterization of mixing

2.2.2.1. Concentration distribution

A piece of information for characterizing the mixing state is the concentration volume distribution. It is a mathematical function whose value at point χ corresponds to the sum of the volumes whose concentration is actually χ. The definition and construction of the concentration volume distribution are explained graphically in Figure 2.1.

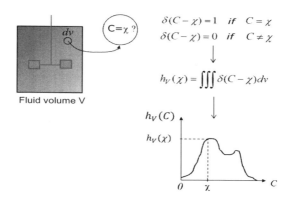

Figure 2.1. *Construction of a volume concentration distribution*

The volume concentration distribution characterizes the homogeneity of the reaction volume (and, in contrast, its heterogeneity). By dividing the volume distribution by the total volume, we get the probability $p(\chi)$ of finding a volume element at a given concentration χ in volume V. Note that the total volume can be seen as the sum of all elementary volumes whatever the concentration that is actually present:

$$\int h_V(\chi)d\chi = V \Rightarrow p(\chi) = \frac{h_V(\chi)}{\int h_V(\chi)d\chi} \qquad [2.9]$$

with the following characteristic properties

$$\int p(\chi)d\chi = 1$$
$$\int \chi p(\chi)d\chi = \langle C \rangle \qquad [2.10]$$

A perfectly homogeneous reactor is therefore characterized by a concentration distribution function equal to a Dirac function centered on the average concentration: $p_\infty(\chi) = \delta(\chi - \langle C \rangle)$.

The concentration distribution gives us purely statistical information. It does not say anything about the concentration's spatial distribution. However, this statistical information is sufficient to calculate the apparent reaction rate on the scale of volume V even if it is heterogeneous:

$$\langle r_C \rangle = \int_0^\infty p(C).r(C)\,dC \qquad\qquad [2.11]$$

2.2.2.2. Concentration field

In a well macromixed reactor, the average concentration is the same at all points of the reactor. A poorly macromixed reactor has spatial concentration differences. If this reactor is compartmentalized into several zones, it may be noted that some have a higher concentration than others. In the case of bioreactors, we think in particular of the zone situated near the feeding point.

From a practical point of view, we can deduce the concentration volume distribution from the knowledge of the concentration field $C(x, y, z)$. However, Figure 2.2 shows that the same concentration distribution can be obtained from different concentration fields, indicating that the concentration distribution contains only part of the information useful for describing, analyzing or modeling a reactor.

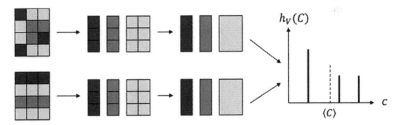

Figure 2.2. *The same concentration volume distribution may correspond to two distinct concentration fields. For a color version of this figure, see www.iste.co.uk/morchain/bioreactor.zip*

Another piece of information is contained in the concentration field: this is the characteristic size of the fluid packets of the same concentration. This information is obtained by analyzing the spatial distribution of the fluid packets through a function known as spatial autocorrelation, which takes the following form:

$$f(l) = \langle C(x)C(x+l) \rangle \qquad\qquad [2.12]$$

Let us consider a point x in the reactor, a second point situated at a distance l from the first one, and calculate the product of the concentrations present at these points. It is found that the product will have a maximum value if the two concentrations are identical. Repeat this for each point in the domain and average the results. The final value indicates the probability of finding, at a distance l, a fluid volume having the same concentration as the reference volume.

Let us focus on the concentration fields shown in Figure 2.3. We see that, this time, the autocorrelation function varies from one concentration field to the other. For field (a), whatever the chosen reference point, it is quite probable to find a point at the same concentration as long as the length l remains less than the distance $l_1 = 3$. Beyond this distance, the probability is practically zero. For field (b), the probability is high if $l < l_1 = 1$, it decreases when $l_1 < l < l_2$ and then drops without becoming completely nil (we can form some pairs of points distant from $l = 4$ which have identical concentrations).

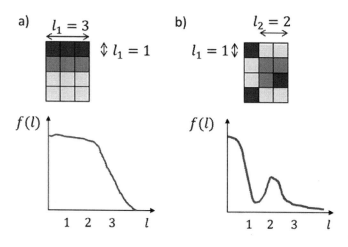

Figure 2.3. *Concentration field, characteristic scales and spatial autocorrelation. For a color version of this figure, see www.iste.co.uk/morchain/bioreactor.zip*

The spatial autocorrelation function of the concentration field thus reveals the structure of the latter. The first part of the curve gives us the characteristic size of coherent fluid packets, and the second part with its optional peaks gives an indication related to the average distance between these fluid packets. By way of illustration, a suspension of droplets can be considered, and two characteristic dimensions appear in the spatial autocorrelation function: the average size of droplets and the average distance between them.

We can still make some useful observations for the following:

– The analysis of a concentration field implies, first, the definition of a resolution scale, Δx below which differences are no longer distinguishable. This does not mean that the concentration is homogeneous under this scale. This means that we access only the average value of the concentration in the volume of dimension Δx^3.

– The modeling of a heterogeneous reactor in concentration also calls for the notion of spatial resolution. In this case, the resolution scale is associated with the hydrodynamic model. The notion of mixing is therefore intimately linked to the notion of observation scale. This explains the use of terms such as macromixing and micromixing to which we will return in the following paragraph.

– In practice, the concentration field can be measured by laser imaging techniques such as PLIF[1] [DAN 07, WAD 05] or calculated using numerical simulation tools by solving the equations of fluid mechanics with or without the consideration of biological reactions [DAN 06, HAR 16, MOR 14].

– In the case of bioreactors, microorganisms travel within the reactor and are therefore repeatedly exposed to concentration variations in their environment. As written in a report by a group of European researchers in the early 2000s: "The behavior of microorganisms is an integrated consequence of concentration fluctuations seen along their trajectory" [ENF 01]. Beyond the knowledge of the concentration field, it is also essential to consider the durations of exposure to different environments either from the residence time of the microorganisms in the different zones of the reactor [BAJ 82], or from numerically simulated velocity fields [LAP 06, MOR 14].

2.2.2.3. *Macromixing and micromixing*

The notion of "good macromixing" or "macromixed reactor" corresponds to the idea we used in Chapter 1 to establish the equations at the reactor scale assuming volume fractions and homogeneous concentrations throughout the volume. The term micromixing introduces an additional scale concept illustrated in Figure 2.4. If we imagine that the black and white circles

---

1 Planar laser-induced fluorescence.

represent different fluid packets, the first situation on the left corresponds to a heterogeneous reactor in concentration at the reactor scale. There are areas with higher concentrations in one of the two reactants. The second situation, in the center, corresponds to a homogeneous spatial distribution of the packets without mixing the packets together. In this situation, the apparent reaction velocity is controlled by the mixing rate at small scales, Vmm. Micromixing thus corresponds to the notion of effective encounter between molecules. The third situation, on the right, corresponds to packets distributed homogeneously and intimately mixed with each other. As we shall see later, the interaction between mixing and reaction will deeply question these notions of characteristic size and time of physical and biological phenomena.

Figure 2.4 illustrates this situation in the case of a second-order reaction.

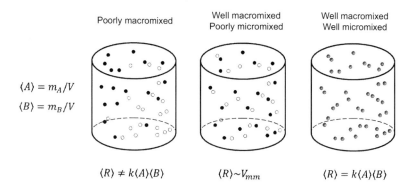

Figure 2.4. *The apparent velocity of a reaction of order ≠ 1 (in this case order = 2) depends on the state of mixing of the reactor (adapted from Fox [FOX 03])*

To illustrate these notions, we can imagine a reactor equipped with four probes measuring the concentration over time. From each time record, a local concentration distribution can be constructed (Figure 2.5). The ordinate $p(C)$ corresponds to the number of measurements where the concentration is C divided by the total number of measurements. It is clear that sufficient data must be acquired for the distribution to be representative (statistically converged).

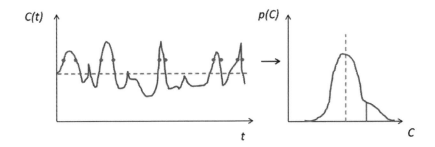

Figure 2.5. *Formation of the local distribution of concentration
from the temporal recording at one point*

We can now get an idea of what local concentration distributions would look like in the cases presented in Figure 2.4. Figure 2.6 shows the concentration distributions in the four zones of a reactor with a macroscopic and/or microscopic mixing defect. In a well macromixed reactor, there is no difference in mean concentration between the different zones (noted from 1 to 4). If it is micromixed, the distribution of concentration at each point is very narrow. If it is poorly micromixed, the mean concentration is the same but the local distribution varies from one zone to another, it is also more spread out. Finally, in a poorly macromixed reactor, the mean concentration varies from one zone to another and local distributions may present a multimodal aspect. This means that the probe measures either a very high concentration of one of the products or a very low concentration.

In modeling, a well-macromixed and well-micromixed reactor can logically be represented using a 0-D approach: a single concentration is sufficient to define the reactor's state. A well-macromixed but poorly micromixed reactor can be approached using a single equation dealing with the evolution of the concentration distribution function. Finally, the modeling of a poorly macromixed reactor requires the association of a hydrodynamic model and a model describing the concentration heterogeneity in each zone. Rodney Fox, in his work on the modeling of turbulent reacting flows, explains that a poorly macromixed reactor is also necessarily poorly micromixed due to the transport of concentration fluctuations due to the turbulent movement between the different zones [FOX 03]. This implies that it will be necessary to provide the means of transporting the distribution function between the various zones of the reactor. We will see some practical examples at the end of this chapter and we will also return to this in Chapter 4 on population balances.

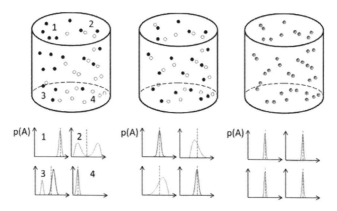

Figure 2.6. *Schematic representation of the concentration volume distributions at different points of a reactor as a function of the micromixing state (according to Delafosse [DEL 08])*

2.2.2.4. *Experimental illustration*

Figure 2.7 illustrates a situation of poor macromixing in an 8 m^3 industrial bioreactor. The mean concentration is not uniform at different points in the reactor. When the injection is located at the bottom (point i1), the mean substrate concentration at point p1, located at the bottom of the reactor, oscillates around 15 mg.l^{-1}. The mid-height and top concentrations oscillate around the mean value of 5 mg.l^{-1}. There is a macroscopic concentration gradient at the reactor scale. When the substrate is injected at the top of the reactor, spatial variations in the mean concentration are still observed. It should be noted, however, that the concentrations measured at the top of the reactor are this time much higher, of the order of g.l^{-1}. This is explained by the fact that the mixing intensity at the injection point is very high for point i1 situated opposite to the stirring turbine, whereas this intensity is markedly lower for point i2 situated at the top, behind a baffle. A further analysis based on more frequent samplings also shows that the structure of the time fluctuations of the concentration depends on the point of measurement: points p4 and p5 are located at the same level but at 90° from each other and on the other side of the baffle.

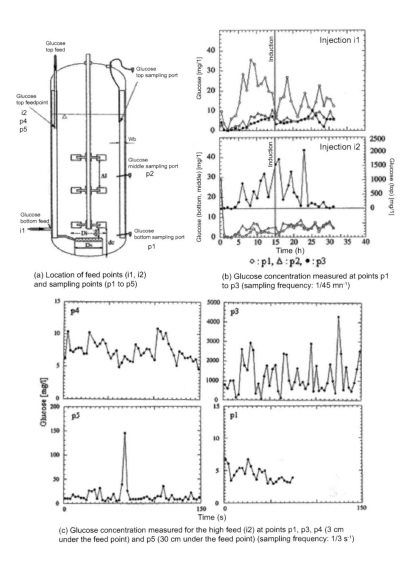

(a) Location of feed points (i1, i2) and sampling points (p1 to p5)

(b) Glucose concentration measured at points p1 to p3 (sampling frequency: 1/45 mn⁻¹)

(c) Glucose concentration measured for the high feed (i2) at points p1, p3, p4 (3 cm under the feed point) and p5 (30 cm under the feed point) (sampling frequency: 1/3 s⁻¹)

Figure 2.7. *Spatial heterogeneities of carbonated substrate concentration measured at different heights and for two feed positions in an 8 m³ bioreactor [BYL 98, BYL 99]. Average concentrations vary from one point to another (bad macromixing). At the same point, significant temporal fluctuations are visible (transport of concentration fluctuations)*

2.2.2.5. Rate of mixing

We have just seen that, in a poorly mixed medium, a concentration distribution exists. Mixing at the different scales modifies this distribution by making it tend towards a Dirac distribution centered on the mean concentration. The difference between the concentration distribution and the Dirac function centered on the mean of this distribution makes it possible to characterize the heterogeneity of the reaction volume. The larger the difference, the greater the likelihood of finding a volume at a concentration that is different from the mean. Since the difference can be positive or negative, the quantification of the mixing state is done by calculating the sum of the squared deviations weighted by the probability that this difference exists. Mathematically, this is the second-order moment of the centered distribution. This characteristic quantity of the distribution is called variance. Details are given in Figure 2.8.

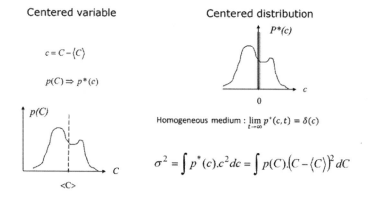

Figure 2.8. Characterization of the mixing state: variance of the distribution

We can thus write that mixing is a set of physical mechanisms that reduce the variance in the concentration distribution. We then define a mixing index (in practice called a segregation index) as the ratio between the initial variance and the variance at time t. This index evolves between 1 initially and 0 when the mixture is homogeneous.

The evolution over time of the segregation index was used in the PhD work of Angélique Delafosse to measure the mixing efficiency as a function of the feed point location. Numerical tracer experiments were conducted along with the large eddy simulation of the fluid flow in a 70 l stirred tank [DEL 08, DEL 09]. By dividing the volume of the vat into 1.2 million elementary volumes, we accurately get the concentration distribution and its variance.

$$C_V = \frac{m_o}{V}$$

$$\sigma^2(t) = \frac{1}{V} \sum_{i=1}^{N_C} \left(C_i(t) - C_V \right)^2 v_i$$

$$I_s(t) = \frac{\sigma^2(t)}{\sigma^2(0)}$$

[2.13]

– m_0 corresponds to the initial mass of the injected tracer;

– V is the fluid's volume;

– $C_i(t)$ is the concentration in the elementary volume v_i;

– C_V is the mean concentration in the vat (constant).

The results obtained during the first few seconds of mixing are shown in Figure 2.9. It is observed that the segregation index decreases rapidly in all cases but that the rate of mixing is greater for an injection carried out at point B, in the jet zone created by the radial stirrer. This is the area where energy dissipation is greatest. The link between the turbulent dispersion and the local rate of mixing is thus illustrated here.

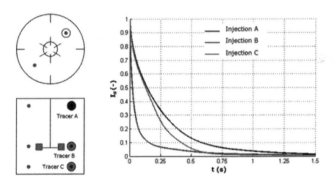

Figure 2.9. *Evolution of the segregation index in a stirred vat of 70 l equipped with a Rushton turbine. The effect of the position of the injection point on the mixing speed is studied by numerical simulation LES [DEL 08]. For a color version of this figure, see www.iste.co.uk/morchain/bioreactor.zip*

The initial rate of mixing corresponds to the slope of the segregation index curve. Hence, a characteristic time of mixing can be defined in the following manner:

$$\frac{d\sigma^2}{dt} = \frac{1}{\tau_m}\sigma^2 \qquad\qquad [2.14]$$

It is essential to understand that these curves describe the first instants of mixing. They therefore show the rate of mixing at the injection point. The worst choice from the point of view of the tracer's dispersion would in this case be to inject it at the top of the reactor. An injection at the top of the reactor tends to produce an area of high segregation in the upper part of the reactor. This reveals the experimental results presented in Figure 2.7. We can also relate this discussion to Dunlop's experimental results [DUN 90] or to those of Hansford and Humphrey [HAN 66] which demonstrate an effect of the injection point's location on the biological reaction.

NOTE.– The homogeneity at the reactor scale is in all cases obtained after about 15 seconds under the study's conditions (N = 150 revolutions per minute). This time corresponds to what is commonly called the reactor's mixing time. It is a macroscopic mixing time (at the reactor's scale).

2.2.3. *Mixing mechanisms*

This part is based on the notions presented in the works of Baldyga and Bourne [BAL 03] and Fox [FOX 03] applied to the case of bioreactors.

The homogenization of the substrate in the medium is ensured by three mixing mechanisms that occur simultaneously: macromixing, mesomixing and micromixing which can still be subdivided into micromixing by incorporation and micromixing by diffusion. These simultaneous mechanisms are illustrated in Figure 2.11.

1) Macromixing is induced by the convective movements generated by the stirring system. It corresponds to the transport without deformation of the fresh substrate packets fed into the reactor at the feeding point. It is a function of numerous parameters including the reactor's geometry (presence or absence of dead volumes), the rotation speed, as well as the type of impeller.

2) Mesomixing is responsible for the fragmentation and size reduction of the fresh substrate packets. This fragmentation is controlled by turbulence. Each vortex present in the flow contributes, on its scale, to dispersing the substrate. It is observed that the decrease in the concentration variance precisely follows the decrease in the velocity fluctuations.

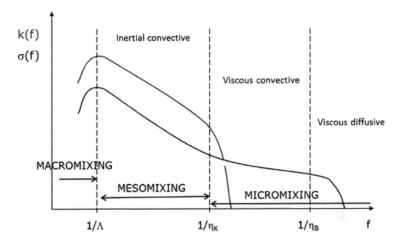

Figure 2.10. *The spectra of the turbulent kinetic energy k and the concentration variance σ as a function of the wavelength f = 1/vortex size. For a color version of this figure, see www.iste.co.uk/morchain/bioreactor.zip*

3) Micromixing (inertial-convective type) performs a winding of the substrate threads formed by mesomixing by giving them a lamellar structure. Micromixing by diffusion dominates from the moment when the fluid's inertia becomes small compared with the viscous friction. The so-called Kolmogorov length scale provides a characteristic size of these fluid volumes. This scale is built from the dissipation of turbulent kinetic energy and the fluid's properties.

4) Micromixing by diffusion is used to homogenize the fluid on a molecular scale when transport by diffusion becomes dominant with respect to the fluid's winding motion described above. The characteristic dimension is called Batchelor's scale.

The power input by unit of mass or the turbulent kinetic energy dissipation rate is an essential parameter to quantify the mixing dynamics. The power input depends on the speed, the stirrer and the density of the liquid. The spatial distribution of this power is very heterogeneous in a stirred tank; the value near the turbine is up to 100 times higher than the mean value [DEL 08, HUC 09].

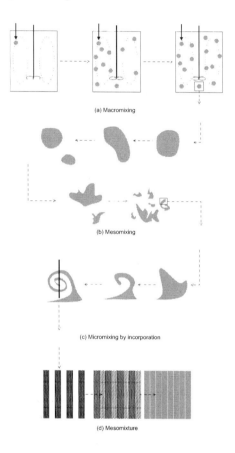

(a) Macromixing

(b) Mesomixing

(c) Micromixing by incorporation

(d) Mesomixture

Figure 2.11. *Mixing mechanisms in a turbulent flow (adapted from Baldyga and Bourne [BAL 03] by A. Delafosse [DEL 08] and M. Linkès [LIN 12b]). For a color version of this figure, see www.iste.co.uk/morchain/bioreactor.zip*

Power input in industrial reactors (ranging from 1 to 10 $kW.m^{-3}$) leads to Kolmogorov scales in the range of 10–50 μm. The size of microorganisms varies from 1 to 10 μm and is smaller or comparable to the Kolmogorov scales, which implies that the substrate brought to a point of the reactor will only be consumable by the microorganisms after micromixing has taken place; as a result, micromixing is solely responsible for the homogenization of the substrate at the microorganism scale.

2.2.4. Characteristic mixing times

Each of the mixing mechanisms acts on a different spatial scale and can be associated with a characteristic time. The temporal and spatial scales are presented in the following table.

	Time	Length
Macromixing	$t_C = \dfrac{V}{N_{Qc} \cdot N \cdot d^3}$	$V^{1/3}$
Mesomixing	$t_S = 2\left(\dfrac{L_C^{\,2}}{\varepsilon}\right)^{\frac{1}{3}}$	$\Lambda = \dfrac{1}{2}\dfrac{k^{\frac{3}{2}}}{\varepsilon}$
Micromixing by incorporation	$t_K = \left(\dfrac{\nu}{\varepsilon}\right)^{\frac{1}{2}}$	$\lambda_K = \left(\dfrac{\nu^3}{\varepsilon}\right)^{1/4}$
Micromixing by diffusion	$t_D = 2\left(\dfrac{\nu}{\varepsilon}\right)^{\frac{1}{2}} . \mathrm{arcsinh}\left(0.05\dfrac{\nu}{D}\right)$	$\lambda_B = \left(\dfrac{D^2\nu}{\varepsilon}\right)^{1/4}$

Table 2.1. *Time and spatial scales of the different mixing mechanisms [DEL 08]. Given the power used in mechanically stirred bioreactors ($\cong$1–10 kW.m^{-3}), cells have a size smaller than Kolmogorov's scale λ_K (10–50 µm) and close to Batchelor's scale λ_B (1–5 µm)*

Macromixing can be described by two characteristic times, the mixing time and the circulation time. The macroscopic mixing time, t_M, is the time at which the concentration at any point of the reactor reaches a constant value (following the punctual addition of an inert tracer). The circulation time, t_C, corresponds to the average time interval between two passages of a fluid particle in the chosen reference zone (generally the volume swept by the stirrer). These two times are strongly correlated. In a liquid medium and in a stirred vessel in the turbulent regime, they are both inversely proportional to the stirring speed N. In a mechanically stirred reactor (single-phase turbulent flow), the mixing time corresponds to 3–4 times the circulation time[2].

The characteristic time of mesomixing corresponds to the time required to reduce the size of the fresh initial fluid packets, L_C, until reaching Kolmogorov's scale λ_k (that of the smallest vortices responsible for the energy transport according to the laws relating to turbulence). At the scale of the reactor, the dissipation rate of the turbulent kinetic energy can be calculated by the following formula:

$$\varepsilon = \frac{N_p N^3 d_a^5}{V}$$

[2.15]

where

– N_p is the stirrer's power number;

– N is the rotation speed in revolutions per second;

– d_a is the stirrer's diameter.

Thus, the increase in the power input has the effect of:

– reducing the gap between Λ and λ_k which reduces the mesomixing time;

– decreasing λ_k, which has the effect of accelerating diffusion transport, and therefore, making the substrate more readily available for consumption.

2.2.5. Contribution of aeration to macromixing

The contribution of the gas phase to the macromixing was approached in the framework of the thesis of Jean-Christophe Gabelle [GAB 12]. It can be summarized by saying that the injection of gas contributes to a significant reduction in the mixing time, in spite of the reduced power input by the impeller in the aerated regime. This conclusion also applies to non-Newtonian fluids.

---

2 This value depends on the type of stirrer.

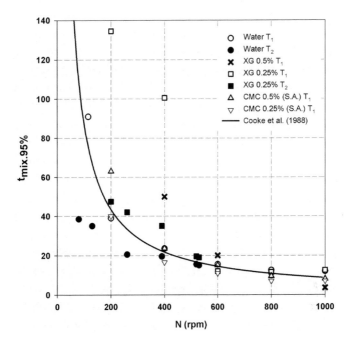

Figure 2.12. *Evolution of a 95% mixing time as a function of stirring speed for Newtonian and non-Newtonian fluids. The symbols refer to experiments carried out with water, xanthan gum or carboxymethylcellulose (source: [GAB 12])*

The data were acquired from two stirred reactors of different sizes (T_1: V=21–42 l; T_2: V=170–340 l) with variable filling levels corresponding to mono- and bi-staged configurations. The reactors are homothetic and are equipped with a Rushton turbine at the bottom and a pitched blade turbine at the top. It is observed that, in the fully turbulent regime zone (at high stirring speed), the measured values practically depend neither on the nature of the fluid nor on the design of the stirring system. The correlation of Cooke *et al.* [COO 88] mentioned in the graph was taken up and extended by Nienow *et al.* [NIE 96].

$$t_M = \frac{K}{N}\left(\frac{P_0}{\rho V}\right)^{-1/3}\left(\frac{D}{T}\right)^{-1/3} T^{2/3}$$
[2.16]

$-\dfrac{P_0}{\rho V}$ represents the power dissipated per unit mass;

$-\dfrac{D}{T}$ is the ratio between the diameter of the stirrer and that of the vessel;

$- K$ is 5.9 for a liquid height equal to the vessel's diameter value of 3.3 which is recommended for multi-stage stirring systems with a liquid height H greater than the tank's diameter T.

It should be noted that for a reactor equipped with several stirrers, the total power corresponds approximately to the sum of the powers due to each stirrer.

When gas is injected into the reactor, it can be seen in Figure 2.13 that the mixing time drops significantly and becomes practically independent of the stirring speed. The gas thus contributes to homogenize the liquid, whatever the rheology of the latter. Two effects add up: on the one hand, the bubbles contribute to the fluid's stirring by dissipating energy in their wake, and on the other hand, the local variations in the gas retention induce large-scale convective movements by a density gradient effect [AUG 03]. Overall, these two contributions decrease when the viscosity increases or when the bubble size decreases. This may explain why there is even a rise in the mixing time for the xanthan concentrated gum at high stirring speed. The beneficial effects of the gas are no longer felt and the adverse effect due to the flooding of the agitators generally leads to a decrease in the power draw compared with the single-phase case at the same speed.

Let us now return to the analysis of the single-phase mixture with viscous fluids. When the stirring rate is lowered below a certain threshold, the values of the mixing time deviate substantially from one another and from Cooke's correlation. This is particularly the case for non-Newtonian fluids. The threshold is however a function of the particular rheology of the fluids in question (rheofluidifiers, for example). Local measurements complementary to the velocity field have shown that the flow can nevertheless maintain a turbulent character in the stirrer's vicinity because of the high shear stresses locally lowering the viscosity of the fluid [GAB 13]. As we move away from the stirrers, the stresses decrease, the viscosity of the fluid increases, and the turbulent character of the flow vanishes. This spatial evolution of the shear rate combined with the nonlinearity of the relationship between viscosity and local shear explains the difficulty of establishing reliable correlations for the mean shear rate in the reactor as a whole. The following correlation is relevant in the highly sheared zone (close to the stirrer):

$$\dot{\gamma}=\left(\frac{\rho N_p d_a^2}{K}\right)^{\frac{1}{n+1}} N^{\frac{3}{n+1}} \qquad with \quad \mu = K\dot{\gamma}^{n-1} \qquad\qquad [2.17]$$

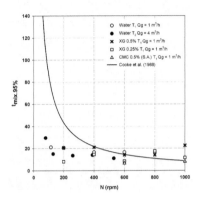

Figure 2.13. *Contribution of the gas phase to the reduction in mechanically stirred vat' mixing time for Newtonian and non-Newtonian media. The symbols refer to experiments carried out with water, xanthan gum or carboxymethylcellulose (source: [GAB 12])*

The Metzner-Otto-type correlation is applied for the calculation of the shear rate far from the stirrer.

Calculation of the average of the shear rate requires knowledge of the volume of each zone, which in practice remains complicated. A computational fluid dynamic calculation makes it possible to estimate these volumes [XIA 09].

Finally, we would like to highlight Professor Alvin Nienow's contribution concerning the effect of shear on the structure of filamentous fungal pellets [AMA 00]. These authors show that the size of the pellets is correlated with the ratio of the energy dissipation to the circulation time. We have just seen that the dissipation occurs essentially in the stirrer's vicinity, *a fortiori* when the fluid shows a rheofluidifying character. Outside this zone, the energy dissipation and therefore the shear are negligible. Moreover, the inverse of the circulation time corresponds to the frequency at which the microorganisms are exposed to intense shear. The proposed relation thus establishes that the morphology of the pellets is related to the maximum amplitude of the stress variation $\gamma_{max} - \gamma_{min}$ multiplied by the characteristic frequency associated with this variation. We will see in Chapter 3 that the metabolic drifts observed in heterogeneous bioreactors can also be related to the maximum amplitude of the concentration fluctuations and the frequency with which the microorganisms are exposed to this change. What could be more natural?

2.2.6. Reaction characteristic time

2.2.6.1. Definition

From a general point of view, the reaction's characteristic time in a homogeneous phase can be identified by linearizing the expression of the kinetic law with respect to the different concentrations. In practice, this amounts to writing the reaction rate in the form of a concentration multiplied by the inverse of a time.

$$r = f(C) \approx \frac{1}{\tau_r} C \qquad\qquad [2.18]$$

For various kinetic laws, we find the following equations:

Kinetic equation	Characteristic time
$r = kC$	$\tau_r = \dfrac{1}{k}$
$r = kC_A C_B$	$\tau_r = \max\left(\dfrac{1}{kC_{A0}}, \dfrac{1}{kC_{B0}}\right)$
$r = kC^n$	$\tau_r = \dfrac{1}{kC_0^{n-1}}$

Table 2.2. *Expression of characteristic reaction times in the homogeneous phase (according to Villermaux [VIL 95] and Balydga and Bourne [BAL 03])*

2.2.6.2. *A case of biological reactions*

While the definition is clear for a chemical system, the application to microbiology poses a number of difficulties:

– The first one concerns the particulate aspect; the cells are not a dissolved species. However, their size is usually small compared with the smallest scales of turbulence and the number of particles per unit volume is very high. In constructing a characteristic time, we may be interested in the overall mass transfer carried out by this very large number of cells and assimilate this sink term to a chemical transformation that would consume the substrate.

– The second concerns the multitude of biological phenomena whose characteristic times cover a very large scale of time (from microseconds to hours) as established by Roels [ROE 82]. It is therefore necessary to precisely define the concerned phenomenon when we evoke a characteristic time for a biological reaction.

Let us illustrate this point by expressing the characteristic time of growth and that of substrate assimilation. For this, we can use a simple, unstructured kinetic model and use equation [2.18] to express the characteristic times:

$$r_X = \mu_{max} \frac{S}{K+S} X \cong \frac{1}{\tau_X} X$$

$$r_S = -q_{max} \frac{S}{K+S} X \cong -\frac{1}{\tau_S} S$$

[2.19]

leading to the following equations:

$$\tau_{R,X} = \frac{1}{\mu_{max}} \frac{K+S}{S} \approx \frac{1}{\mu_{max}}$$

$$\tau_{R,S} = \frac{1}{q_{S\,max}} \frac{K+S}{X} \approx \frac{1}{q_{S\,max}} \frac{S}{X}$$

[2.20]

The characteristic times of assimilation and growth obey different rules: the former does not depend on the cell concentration, whereas the latter is inversely proportional to the cell concentration. Given the maximum growth rate of industrially grown strains ($\mu_{max} < 1\ h^{-1}$), the characteristic time of growth is unconditionally slow compared with all other phenomena within the bioreactor (mixing and transfer between phases). We can say that this is an integral time-scale insofar as growth integrates a whole set of dynamic biological phenomena, the ultimate consequence of which is the formation of new cells. The characteristic time of growth is a property of the considered strain and it is independent of the culture conditions

The characteristic assimilation time, on the other hand, depends on the concentration of cells that consume the substrate. It therefore depends on the operating conditions and can be very short in high cell density cultures. While the characteristic time of growth is specific to the strain, that of substrate assimilation is a property of the cultivation and it evolves considerably during the course of an experiment, particularly in batch and fed-batch cultures because of the accumulation of cells in the reactor.

It should be kept in mind that:

– if μ_{max} allows us to estimate the growth characteristic time, q_{Smax} is not sufficient to calculate the assimilation characteristic time;

– if, at equilibrium (steady state), μ_{max} and q_{Smax} are of the same order of magnitude (with a Y_{SX} coefficient), the specific assimilation rate of a cell experiencing a limitation may temporarily be up to 10 times greater [LAR 09, NEU 95]. We shall return to this point in Chapter 3;

– two continuous cultures performed at the same dilution rate are similar from the point of view of the growth characteristic time, but they are not necessarily similar from the point of view of assimilation. In addition, the cell concentration must be the same, which implies the same concentration of sugar in the feed. As the volume flow rate and the feed concentration are independent, it is possible to realize continuous cultures at the same dilution rate having different assimilation times. From then on, the competition between mixing and assimilation will be more or less severe. The disparity in the values of the triplet (μ_{max}, K_S, Y_{SX}) identified from experimental data uncovers the beginning of a physical explanation.

2.3. Interaction between mixing and bioreaction

2.3.1. *Competition between mixing and chemical reaction*

We will briefly consider this question in order to highlight only a few facts and results that will explain the rest of this section.

The mixing of species precedes the reaction; the two phenomena being consecutive, the dynamics of the whole is imposed by the dynamics of the slowest phenomenon. We thus distinguish two regimes:

– the physical regime in which the rate of transformation is imposed by the rate of mixing. This regime corresponds to a characteristic time of the mixing that is greater than the reaction's characteristic time;

– the chemical regime in which the transformation rate of the reactants corresponds to the actual chemical reaction's rate. This regime corresponds to a mixing time which is less than the reaction's characteristic time.

It is therefore sufficient, in principle, to compare the characteristic times of the two phenomena to identify whether or not the mixing limits the reaction rate.

From a practical point of view, in chemistry, the study of the interaction between mixing and reaction is carried out using a schematic diagram of competitive parallel reactions [BAL 97, FOU 96, VIL 95]. The first reaction is characterized by a very large kinetic constant relative to the second one; in principle, this first reaction is therefore very rapid and the production of the by-product Q by the second reaction, referred to as the side reaction, is extremely small[3].

$$A + B \xrightarrow{k_1} P$$
$$A + P \xrightarrow{k_2} Q$$

[2.21]

Reactants A and B are assumed to be supplied in stoichiometric quantities. If the mixing is perfect, each molecule of A is in the vicinity of

---

3 It depends strictly on the ratio k_2/k_1.

a molecule of B with which it reacts to give the product P and no reactant A remains to react again with the product P. If the mixing is not perfect, a local excess of A may exist and therefore some reactant A will remain in the vicinity of a point where reagent P has just been formed. This leads to the formation of the by-product Q by the second reaction. This side reaction is only possible because reactant B has been locally depleted. The amount of by-product resulting from the side reaction therefore depends directly on the rate of mixing. It is observed that the mixing interferes with the reaction when the mixing time exceeds 1/10 of the main reaction's characteristic time. Thus, the experimental measurement of the Q/P concentration ratio provides an indication of the quality of mixing in the reactor.

A practical methodology used to highlight a competition between mixing and reaction was proposed by the Polish researcher Jerzy Baldyga and the American researcher John Bourne [BAL 03]. We can extract some simple rules:

– the position of the injection point is important because higher values of the turbulent kinetic energy dissipation rate lower the mesomixing time;

– by multiplying the number of injection points, the flow rate is reduced at the level of each injection, which favors the dispersion with respect to convection in the jet formed at the injection.

The modeling of this interaction between mixing and chemical reaction was at first semi-empirical and allowed the results to be explained qualitatively. By integrating the idea that turbulence is the underlying mechanism of mixing at small scales, the parameters of the first models received a more precise quantification in relation to the operating conditions.

2.3.2. Competition between mixing and biological reaction

The analysis of this question is not easy. Unfortunately, an experimental technique as reliable and precise as the one present in chemistry is not available in microbiology. Indeed, the speed of the chemical reaction is entirely defined by thermodynamics. On the other hand, in microbiology, the rate of reactions is also a function of the state of the cells, which is affected by the state of mixing. Moreover, the manifestation of a link between the rate of mixing and the rate of biological reaction depends on the observation scale. Thus, in a continuous reactor, competition between mixing and assimilation will have no effect on the growth rate (imposed by the dilution rate), but more reliably on the conversion efficiency of sugar into cells [YE 85].

In any case, the comparison of mixing times and assimilation times is a rich topic. Here are the numbered and commented conclusions, taken from the thesis of Angélique Delafosse [DEL 08].

	$\varepsilon_V/10$	ε_V	$10\varepsilon_V$
Λ (m)	0.14	0.14	0.14
τ_C (s)	13	13	13
τ_{meso} (s)	0.4	0.2	0.1

Table 2.3. *Estimation of the spatial and temporal scales of macromixing and mesomixing in an industrial bioreactor (20 m³) ($\varepsilon_V = 1.3kW.m^{-3}$)*

	$\varepsilon_V/10$	ε_V	$10\varepsilon_V$
Λ (m)	0.01	0.01	0.01
τ_C (s)	0.8	0.8	0.8
τ_{meso} (s)	0.07	0.03	0.02

Table 2.4. *Estimation of the spatial and temporal scales of macromixing and mesomixing in a laboratory bioreactor (3 l) ($\varepsilon_V = 1.3kW.m^{-3}$)*

	$\varepsilon_V/10$	ε_V	$10\varepsilon_V$
η_k (μm)	50	30	20
τ_{micro} (ms)	50	15	5
η_B (μm)	1.7	0.9	0.5
τ_{microD} (ms)	26	8	3

Table 2.5. *Estimation of spatial and temporal scales of micromixing ($\varepsilon_V = 1.3\ kW.m^{-3}$)*

		$X(g.l^{-1})$					
		5	10	20	30	40	50
$S\ (mg.l^{-1})$	5	1.5	0.8	0.4	0.3	0.2	0.1
	10	3	1.5	0.8	0.5	0.4	0.3
	20	6	3	1.5	1	0.8	0.6
	30	9	4.5	2.3	1.5	1.1	0.9
	40	12	6	3	2	1.5	1.2
	50	15	7.5	3.8	2.5	1.9	1.5

Table 2.6. *Time characteristic values of the assimilation of substrate $\tau_{R,S}$ in seconds. Values calculated from the maximum assimilation rate of S. cerevisiae in batch (q_{Smax}=2.4 $g_S.g_X^{-1}.h^{-1}$)*

$-\tau_{R,S} > \tau_C$: the biological reaction is slow with respect to the macromixing. The other mixing mechanisms being faster than macromixing, the reactor is perfectly macromixed and perfectly micromixed. In other words, there is no concentration heterogeneity, whatever the scale is. This situation occurs especially when the cell concentration is low and the substrate concentration is high. It can be noted that if this situation is the norm in a small bioreactor, in an industrial bioreactor, the probability that the macromixing has no influence is very unlikely.

– $\tau_{meso} < \tau_R < \tau_C$: the biological reaction is fast with respect to macromixing but slow with respect to mesomixing. This implies that substrate concentration gradients appear at the reactor scale. The reactor is poorly macromixed and there are regions of different concentrations, but each region is well micromixed. During their displacement in the reactor, the cells are thus subjected to concentration variations in their environment with an average frequency of the order of $1/\tau_C$. In certain cases, especially if the cell concentration is high, there may be a depletion of the substrate in the time interval between two passages in the feeding zone. Such a situation is typical of industrial bioreactors in fed-batch operation. From the point of view of the representation of the bioreactor, this situation leads naturally to a compartmentalization of the reaction volume in zones of different concentrations [PIG 15, VRÀ 01].

– $\tau_{micro} < \tau_R < \tau_{meso}$: the biological reaction is rapid compared with the time required to reduce the integral concentration scale. Schematically, during their movement in the reactor, the cells' environment will be replenished with fresh fluid at a frequency of the order of $1/\tau_{meso}$. However, the mixing down to the micro-scale of the cells remains faster than the reaction. It may be noted that the initial size of the injected fluid packets is not the only cause. The local characteristics of the turbulence "seen" by the fresh fluid packs along their trajectory are also to be taken into account [FOX 03, KRU 96]. Such a situation is likely to be present in industrial-size bioreactors because the integral scales are directly related to the geometry and vessel's volume. The size of the fresh fluid packets injected thus tends to increase with the size of the reactor. This size also depends on turbulence at the injection point. Hansford and Humphrey's precursor work on a laboratory reactor of a few liters demonstrates the effect of injection position on the efficiency of substrate conversion to biomass [HAN 66].

– $\tau_{R,S} < \tau_{micro}$: the reaction is rapid with respect to the micromixing rate. The direct environment of the microorganisms can be depleted over a period corresponding to the lifetime of the vortices. In this case, the concentration in the vicinity of the cell may be substantially different from the average concentration in the liquid.

$-\tau_{R,S} < \tau_{microD}$: the rate of the biological reaction is limited by the external diffusion flux. In view of the reaction times, this situation does not appear to be possible with the dissipated power values generally encountered in bioreactors, apart from a very low substrate concentration and a high concentration of biomass at the end of the culture, it is unlikely that mixing by diffusion at the cell scale is the cause of a limitation of the assimilation rate.

This analysis is confirmed by the experimental work of Dunlop and Ye [DUN 90], which shows that the substrate-into-biomass yield and the substrate uptake rate depend on the flow rate at the feeding point and the location of the feeding point [AMA 01, LEE 82]. As expected, the yield and uptake rate vary during a fed-batch reactor culture as cell concentration increases [LIN 00, LIN 01].

2.3.3. *Spatial approach to the micromixing problem*

Thus far, we have examined the problem from the temporal point of view by comparing the mixing time and the reaction time by considering the biological reaction as a homogeneous catalysis. This question can also be examined from a spatial point of view by attempting to represent, at least schematically, the concentration field at the biological particle scale. This point of view leads us to question the notion of concentration seen by the cell in relation to that measured on the macroscopic scale (the only practically measurable one). This is an essential issue and largely a matter of heterogeneous catalysis. We shall therefore begin with a brief description of the classical notions used in this field before expanding to the case of biological systems.

2.3.3.1. *A few words on heterogeneous catalysis*

One question posed in this field consists of predicting the macroscopic reaction rate resulting from a chemical transformation performed within or on the surface of suspended particles. In a solid particle, the transport of chemical compounds takes place under the effect of diffusion. Thus, in the case of a spherical particle, we can establish the conservation equation of a constituent and the associated boundary conditions (central symmetry and continuity of fluxes at the interface) in the form:

$$\frac{\partial C(r,t)}{\partial t} - \frac{1}{r^2}\frac{\partial}{\partial r}\left(r^2 D \frac{\partial C(r,t)}{\partial r}\right) = P(C(r,t)) \quad r \in \,]0, R[$$

$$\left.\frac{\partial C(r,t)}{\partial r}\right|_{r=0} = 0 \qquad\qquad\qquad [2.22]$$

$$-D\left.\frac{\partial C(R,t)}{\partial r}\right|_{r=R} = k_L \left(C(R,t) - C_\infty\right)$$

where:

– R is the particle's radius;

– D is the diffusion coefficient in the particle;

– k_L is an external transfer coefficient, whose expression integrates the properties of the fluid, the particle and the flow regime;

– C_∞ is the concentration far from the particle (that which will be measured in practice);

– $P(C)$ is the source (production) term, here the expression of the reaction rate.

By solving this equation, we obtain the concentration field in the particle and the volume integration of the reaction term then makes it possible to express the reaction rate, <P(C)>, per unit mass of the particle [2.23]. On the macroscopic scale, the overall apparent velocity is then written as the product of this specific velocity by the total mass of particles. However, this approach is relatively heavy, even for a steady state. It will be assumed here that the possible variations in C_∞ are slow enough to allow the concentration fields to be established.

$$\langle P \rangle = \frac{4\pi}{\rho_p V_p} \int_0^R P(C(r))r^2 dr \qquad\qquad [2.23]$$

The aim of heterogeneous catalysis is therefore to express the overall apparent velocity from macroscopic physical quantities, in particular C_∞, k_L and D.

For this purpose, two efficiencies are introduced: the internal efficiency which aims to account for the catalyst grain's homogeneity and the external efficiency which integrates the effect of the gap between the concentration in the liquid and that at the particle's surface.

$$\eta_i = \frac{\langle P \rangle}{P(C(R))}$$

$$= \frac{observed \;\; reaction \;\; rate}{reaction \;\; rate \;\; based \;\; on \;\; the \;\; concentration \;\; at \;\; the \;\; surface}$$

$$\eta_e = \frac{\langle P \rangle}{P(C_\infty)}$$

$$= \frac{observed \;\; reaction \;\; rate}{reaction \;\; rate \;\; based \;\; on \;\; the \;\; concentration \;\; far \;\; from \;\; the \;\; surface}$$

[2.2]

Indeed, we can distinguish several asymptotic cases in which the problem can be described in very simple terms:

– If the transport by diffusion within the particle is very rapid in comparison with the reaction rate, the grain is practically homogeneous in concentration. The entire volume of the grain is efficient because the reaction occurs. The internal efficiency is then 1 and the reaction rate per unit volume is written as:

$$\langle P \rangle = \frac{4\pi}{\rho_p V_p} \int_0^R P(C(r))r^2 dr = P(C)$$

[2.25]

The complete solution of the problem is obtained by writing that the total mass reacted in the particle per unit time is equal to the mass flux transferred to the interface, namely:

$$P(C)\rho_p Vp = k_L \left(C - C_\infty \right) \pi R^2$$

[2.26]

– If the rate of transport of the constituents to the particle is very rapid in comparison to the reaction rate in the whole grain, the concentration at the particle's surface is practically equal to the concentration measured in the liquid. In this case, the external efficiency is 1 and the solution of the problem can be calculated using [2.22] assuming that $C(R) \approx C_\infty$.

– If the diffusive transport is slow in comparison with the reaction rate, the grain is heterogeneous in concentration. We can always, after having solved problem [2.22] for a given kinetic law (order 1 or 2), integrate the local reaction rate with the particle's volume and thus obtain the mean velocity per unit volume, but the link with the concentration C_∞ is not direct anymore. We then use the notion of internal efficiency whose role is to correct the value based on the hypothesis of a homogeneous grain.

– In the same way, if the speed of the external transport is of the same order of magnitude as the reaction rate at the particle scale, the concentration at the liquid–solid interface differs from C_∞. We then introduce a second correlation, the external efficiency, the aim of which is to correct the value which would be obtained by making the hypothesis of a homogeneous liquid up to the particle's scale.

From a practical point of view, we will therefore try to evaluate the ratio between rates or characteristic times of the reaction and the internal and external transport – we will thus introduce Damkhöler and Biot dimensionless numbers; then *correlations* will be used which define the efficiency values as a function of these dimensionless numbers. We must emphasize the use of the term correlation: it refers to a synthesis of results which is derived here from fine modeling at the particle's scale. Too often it is thought that a correlation is essentially experimental. In the framework of multi-scale modeling, a correlation is a law which makes it possible to account, on a given scale and in an integrated way, for phenomena occurring at finer unresolved scales.

2.3.3.2. *Transposition to the case of biological particles*

The case of an enzymatic reaction carried out on the surface of functionalized support particles is quite identical in principle to the cases studied previously. The notion of internal efficiency does not need to be and it suffices to consider a Damkhöler number based on the external transport and the surface reaction rates.

In the thesis work of Ye, we find an initial attempt to relate the stirring intensity to the biological reaction's rate, considering that the cells of radius r_c are at the center of a fluid sphere of radius R_∞, inside which transport is controlled by diffusion. The starting point is to relate the assimilation term to the transfer term per unit volume:

$$\phi_{LS} = v_{\max} \frac{C}{K+C} X = k_L a (C - C_\infty) \qquad [2.27]$$

By solving the diffusion scalar transport equation in the area surrounding the cell, it is established that:

$$\phi_{LS} = \frac{6D(C_\infty - C)}{R_\infty^2 - r_C^2}$$ [2.28]

From this, we derive the value of the interfacial concentration in the case where it is small in front of the affinity constant K:

$$C = \frac{C_\infty}{1 + \dfrac{v_{max}X}{6DK}\left(R_\infty^2 - r_C^2\right)}$$ [2.29]

Finally, an external efficiency can be defined by comparing the rate of actual assimilation with that built up on the concentration far from the cell:

$$\eta = \frac{r(C)}{r(C_\infty)} = \frac{\dfrac{C}{K+C}}{\dfrac{C_\infty}{K+C_\infty}} = \frac{C}{C_\infty}\cdot\frac{K+C_\infty}{K+C} = \frac{K+C_\infty}{K+C_\infty + \dfrac{v_{max}X}{6DK}\left(R_\infty^2 - r_C^2\right)}$$ [2.30]

By assuming a zero-order reaction rate on the cells' surface and an external diffusion-type transport, Asenjo and Merchuk established the similar expression they use to explain the relationship between the variability of the kinetic constants and the state of mixing [MER 95]. For their part, Linkès and his colleagues performed unsteady numerical simulations by considering concentration fluctuations in the field distant from the particle [LIN 12a]. The size of the fluid domain where diffusion dominates is related to turbulence scales and the assimilation at the interface is dynamically treated according to whether the transfer flux is limiting (zero concentration at the interface) or the cell assimilation capacity is limiting. The results show that the concentration at the cell's surface can be transiently canceled out, whereas the mean (time) flux is non-zero. This provides a plausible explanation for the activation of the biological responses characteristic of famine and starvation, whereas the measured concentrations are non-limiting [GAR 09].

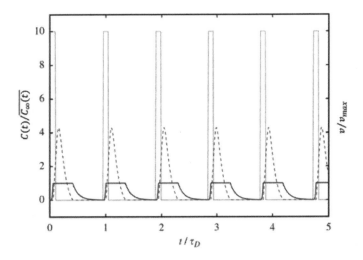

Figure 2.14. *Temporal evolution of the distant field (fine dotted line), the concentration at the particle normalized by the mean distant field concentration (dashed) and the normalized assimilation rate as a function of time (continuous line) (abscissae normalized by the diffusion characteristic time) [LIN 12a]*

2.4. Analysis and modeling of couplings between mixing and bioreaction

The modeling of the reactors refers to two asymptotic cases, the continuous stirred-tank reactor (RAC) and the plug flow reactor. In the case of a non-ideal reactor, it is necessary to distinguish between hydrodynamics and the mixing state. It is the combination of these two concepts that defines the reactor model completely. It thus becomes judicious to distinguish, for example, a plug flow reactor in a state of perfect mixing and a plug flow reactor in a segregated state (imperfect mixing). Here we refer indeed to a non-premixed plug flow reactor, which is quite explicit. The hydrodynamic model provides a description of the fluid velocity and, if necessary, turbulence, while the mixing model describes the evolution of the state of segregation (or concentration distribution).

To build such a model of a reactor, we can nowadays rely on computational fluid dynamics codes. The reaction volume is then cut into a few million elementary volumes. On each of these volumes, the fluid mechanics equations are solved and, in each volume, a concentration distribution is defined, which evolves due to the convective transport, the local turbulent mixing and the reaction's intensity. However, it is an expensive approach in terms of time and computing resources which requires a very high level of expertise and a mathematical description of the various phenomena, both fine and robust.

Part of the modeling work is therefore to improve the physics of the models in order to use them in simulations that are always solved in a better way. Another part of the work is to produce integrated models, compatible with the means, objectives and knowledge of engineers. In this last part, we will focus on this type of model for two reasons:

– integrated models are pedagogical because they prompt us to reflect on the nature and consequences of simplifications;

– dynamic simulation of a bioreactor using a numerical fluid mechanics approach still requires mathematical developments to manage the large number of variables describing the state of local mixing and the state of the biological system.

2.4.1. *Link between the segregation state and calculation of the apparent rate of a simple biological reaction*

In this first example, we will be interested in the evaluation of the reaction term at the reactor scale in which it will be assumed that there exists a known concentration distribution. Furthermore, the reaction rate as a function of concentration is also known. Finally, it is assumed that the reaction rate depends only on the concentration at the considered point.

We therefore use a 0-D hydrodynamic model, which is of zero spatial dimension, and assume the shape of the concentration distribution that is invariant over time. This vision is suitable for the modeling of a continuously stirred bioreactor (or Chemostat).

To illustrate this, we will consider the case of a bioreactor and in particular the conservation equation of the substrate in the simplest form. In Chapter 1, we established this equation:

$$\frac{\partial S}{\partial t} = \frac{F_i S_i - F_o S}{V} + R_S V \qquad [2.31]$$

The last term corresponds in fact to the resultant of all the local consumption rates realized at different points of the reactor. If we want to be exact, this equation results from the integration on the reactor's volume of a local equation established on an elementary volume dv assumed to be homogeneous. In fact, it describes the evolution of the average substrate concentration which will be noted as <S>:

$$\frac{\partial \langle S \rangle}{\partial t} = \frac{F_i S_i - F_o \langle S \rangle}{V} + \iiint_V r_S dv = \frac{F_i S_i - F_o \langle S \rangle}{V} + \left(\frac{1}{V} \iiint_V r_S dv \right) . V \qquad [2.32]$$

Thus, it can be seen that the rate R_S in fact denotes the volume average of the local rates, $R_S = \langle r(S) \rangle$. The question that arises is how do we relate the average local rates to the mean concentration <S>? Is it right to say that the volume average of the local rates is equal to the rate calculated for the mean concentration? To answer this last question is to identify the conditions under which the following relation would be exact:

$$\langle r(S) \rangle = r\left(\langle S \rangle \right) \qquad [2.33]$$

The general answer to this question which was already addressed at the beginning of this chapter is that the relation is exact only if the kinetics is of order 0 or 1 with respect to the concentration, which means here that $r(S) = k$ or $r(S)=kS$.

A priori, this is not the case for biological kinetics which are under the control of enzymatic reactions for which a Michaelis–Menten-type equation is most often adopted:

$$r(S) = r_{max} \frac{S}{K_S + S} \qquad [2.34]$$

We can observe, however, that this rate law is a bilinear or quasi-bilinear law. For example, other models like that of Blackman are sometimes used [KOC 82]. Therefore, it can be expected that the mixing state will have a small effect on the calculation of the substrate consumption.

To verify this idea, we can examine the results of using the two approaches proposed in equation [2.33] for the reaction rate calculation.

Different situations are illustrated in Figures 2.15 and 2.16. The presumed concentration distribution is plotted in red and the black curve shows the evolution of the reaction rate as a function of the concentration [2.34]. The latter curve is therefore identical for each case. From the concentration distribution, it is easy to calculate the mean concentration and the value of the reaction rate at that concentration. An expression identical to that of equation [2.11] to calculate the exact reaction rate is used.

$$\langle r(S) \rangle = \int_{0}^{\infty} p(S).r(S)\,dS \qquad [2.35]$$

The results are presented in non-dimensional form by taking the ratio S/K_S on the abscissa and the ratio r/r_{max} on the ordinate.

In Figure 2.15(a), the presumed concentration distribution is monomodal and narrow. The difference between the two approaches exists and it can be quantified numerically, but it is so small that it is not necessary to favor one method or another for the calculation of the apparent reaction rate.

In Figure 2.15(b), the presumed concentration distribution is monomodal but much more spread out. It covers the entire nonlinearity zone of the rate law, $0.5 < S/K_S < 4$. In fact, the difference here is slightly more sensitive, but it remains in the order of a few percent. A very precise experimental method should be available to measure the concentration distribution in order to assert that the calculation based on the mean value of S is only an approximation of the exact value. In practice, in biology, means are not available for measuring concentrations with such precision, let alone their distribution.

In Figure 2.16(a), the presumed distribution is bimodal with two peaks located on either side of the nonlinearity zone. The average concentration is very unlikely. Care should be taken not to interpret this distribution as resulting from two compartments at different concentrations. The reacting volume is highly segregated, and there is no indication, for example, that all the low-concentration fluid elements are contiguous (see Figure 2.2).

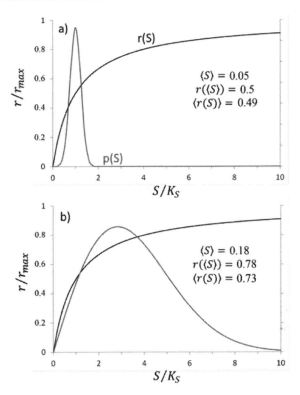

Figure 2.15. *Consequence of the form of the concentration distribution on the calculation of the average growth rate. For a color version of this figure, see www.iste.co.uk/morchain/bioreactor.zip*

The rate difference here becomes appreciable from a calculation point of view and is greater than 10%. What happens in practice? For the experimenter who only has access to the average value of the concentration, of which we assume that he/she will make a perfect measurement, the use of the macroscopic law $r(<S>)$ leads to a value of r_{max}, which only differs by 10% from the actual value. It is therefore sufficient to allow a 10% change in the value of r_{max} in the equation $r(<S>)$ to obtain the exact value of the apparent reaction rate $<r(S)>$. A similar approach, impacting the constant K_S, can be devised. In practice, it is of course a step of adjusting the parameters r_{max} and K_S that will be preferred to match the rate calculated by a macroscopic law to the measured apparent rate. However, the identified values will bear the mark of the existence of a concentration distribution in the reactor. Thus, each segregation state will produce a slightly different set of parameters. It can thus be asserted that the segregation state in a bioreactor will generate noise in the identification of the parameters of the biological law.

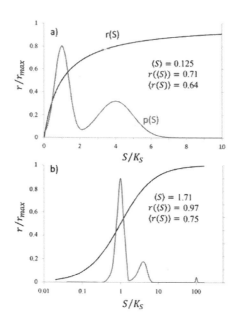

Figure 2.16. *Consequence of the form of the concentration distribution on the calculation of the average growth rate (case of multi-modal distributions). For a color version of this figure, see www.iste.co.uk/morchain/bioreactor.zip*

Owing to the low nonlinearity of kinetic laws, we must have a very marked segregation state in order to observe a significant difference between the two modes of calculating the biological reaction rate. Based on this view, we would then conclude that there is little effect of the segregation state on biological reactions. This is contrary to the numerous experimental observations which affirm an effect of the mixing state on biological kinetics. This is a paradox that will need to consider things a little more in detail.

In this first example, we assumed the existence of a concentration distribution. We have given it some characteristic forms to identify the link between this form, the rate law and the calculation of an apparent rate based on the average concentration or on the complete distribution.

In a real bioreactor, the concentration distribution results from the competition between the mixing and assimilation of the substrate. Unfortunately, the experimental measurement of the glucose concentration distribution, for example, is not accessible. If the state of segregation is important but its measurement is practically out of reach, modeling can be an interesting alternative. In doing so, the contributions of physical origin and those of biological origin will be clearly distinguished and the share of each one will be quantifiable in the final result.

2.4.2. Modeling of non-perfectly mixed bioreactors

The modeling of a reactor is based on the definition of a hydrodynamic model and that of the mixing state.

2.4.2.1. DTS approach and mixing state

The first series of work on the effects of mixing in bioreactors may be attributed to Fan *et al.* [FAN 70, FAN 71, TSA 71]. This work concerns both the identification of the operating conditions likely to lead to the reactor's wash-out, the stability analysis and the theoretical study of several combinations of ideal reactors with or without recycling. Regarding the biological reaction, simple kinetic models are used.

An open reactor is partly characterized by the distribution of residence times $E(\theta)$. This is the probability that a fluid particle stays during a time θ in the reactor. For a plug flow reactor, all particles have the same residence time. The resulting distribution is therefore a Dirac distribution centered on the residence time $\tau = \dfrac{V}{Q}$. In a general case, the distribution of the residence times is a certain function and it is useful to observe that any distribution can be seen as a sum of a Dirac function.

$$x(t) = \int_0^{+\infty} x(\theta)\delta(t-\theta)d\theta \qquad [2.36]$$

From there, and from the point of view of the distribution of the residence times, it is thus possible to represent an actual reactor in the form of a series of plug flow reactors placed in parallel. The length of each reactor determines a particular residence time. The fraction of the total flow rate passing through each reactor determines the probability associated with this particular residence time.

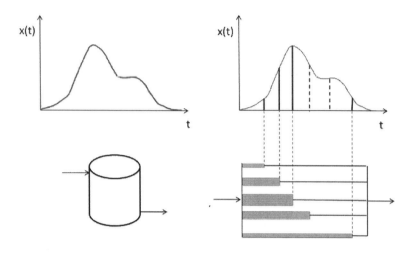

Figure 2.17. *Distribution of residence times in a continuous reactor and representation in the form of a set of parallel threads. The width of each thread corresponds to the fraction of the total flow injected, and the length corresponds to the residence time in this thread*

Under the strong hypothesis of a so-called complete segregation state and for a reaction rate of order 1, it is then possible to determine the concentration at the outlet of a real reactor from the following formula.

$$C_s = \int_0^\infty C(\boldsymbol{\theta})E(\boldsymbol{\theta})d\boldsymbol{\theta} \qquad\qquad [2.37]$$

This formula indicates that the output concentration is the sum of the concentrations $C(\theta)$ weighted by the probability $E(\theta)$ of staying for a time θ in the reactor. $C(\theta)$ results from the integration between t = 0 and t = θ of the kinetic equation established in a closed reactor. For a kinetic of order 1, we will then have:

$$C(\boldsymbol{\theta}) = C_0 e^{-k\theta} \qquad\qquad [2.38]$$

In the case of more complex kinetics, it will be necessary to integrate the system of corresponding differential equations from the initial conditions.

The approach adopted in this type of modeling consists of representing the complete reactor as a combination of ideal reactors (continuous or plug flow reactor) and to define for each one a state of mixing (perfect mixing or complete segregation). The necessary definition of the initial conditions at the inlet of each sub-compartment implies making a new modeling choice concerning the homogenization of concentrations between two compartments [FAN 71]. Thus, these authors show that for two reactors in series, we must already consider a total of six different models.

A similar approach has been proposed by Bajpai and Reuss to describe the interaction between mixing and biological reaction in a heterogeneous bioreactor [BAJ 82]. The idea consists of using the circulation time distribution, that is, the time elapsed between two passages in a particular zone of the reactor. This information is more complex to obtain than the distribution of residence times. The notion of the trajectory of fluid particles is approached a little more, but it is still statistical information. This type of analysis has emerged for semi-closed or continuous reactors when it was found that the frequency of the passage of microorganisms in the substrate's injection zone was decisive. The case studied in the Bajpai and Reuss study can be regarded as the association of a perfectly mixed zone followed by a zone in a state of complete segregation characterized by a residence time distribution and a recycling of this zone towards the perfect mixing zone. This modeling seems more physical but it omits the existence of an exchange between the fluid threads along the trajectories which bring them back to the feeding point.

The limitation of this approach combining the residence time distribution (or circulation) and the state of mixing is threefold:

– The decomposition of the actual reactor into a sub-volume set is not unique and subsequent works have shown that if the reaction is not of order 1, the decisive information concerns the moment when the injected molecules meet effectively. However, this information is essentially absent from the distribution of residence times, which is based on an input–output-type analysis. In fact, combining a plug flow reactor followed by a continuously stirred reactor or the reverse combination produces the same residence time distribution but leads to different conversion rates. How can we select the relevant decomposition?

– The final result depends on the choice of the segregation model in each sub-volume and at each junction between sub-volumes. Combined with the previous decomposition, these choices will determine the segregation index which should be as close as possible to that of the actual reactor.

– The choices can only be validated *a posteriori*, that is to say by showing that such and such combination makes it possible to account for the experimental results.

Being able to explain the observations and provide a solid theoretical basis for analyzing them cannot be considered as a minor objective in modeling. However, we can also have the ambition to carry out a predictive modeling in which the parameters would be defined starting from the operating conditions and no longer adjusted in order to reproduce the observations. To advance in this direction, we must add the information relative to the mixing's dynamics to the model. Of course, it seems tempting to turn to a detailed description of the reactor's internal hydrodynamics using a numerical simulation. We will see, however, that there exist intermediate approaches.

2.4.3. *Approach based on a mixing model*

We will focus here on the case of a non-premixed plug flow reactor. The plug flow reactor is shown schematically in Figure 2.18. In contrast to the ideal plug flow reactor, the incoming streams are not premixed. The concentration is not homogeneous in a cross-section of the reactor. The mean fluid velocity, U, is assumed to be constant, equal to the ratio of the total flow rate to the reactor's cross-section.

Although the concentration $C(r,z)$ depends both on the radial and axial positions, it is sought to establish a model of this one-dimensional (axial) reactor, while taking account of the imperfect mixing state. It will be seen that under certain conditions, the case of a closed reactor having initial concentration heterogeneity can be described by the same set of equations.

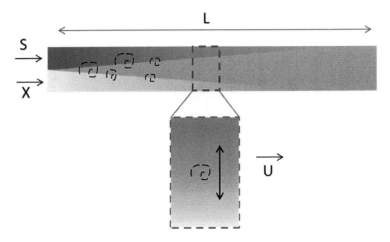

Figure 2.18. *Diagram of a non-premixed plug flow reactor. The spirals symbolize the radial mixing mechanism. For a color version of this figure, see www.iste.co.uk/morchain/bioreactor.zip*

2.4.3.1. Hydrodynamic model 1-D

Let us begin by writing the general equation of transport of the average concentration[4] of a compound C in liquid phase:

$$\frac{\partial \langle C \rangle}{\partial t} + U \frac{\partial \langle C \rangle}{\partial z} - E_z \frac{\partial^2 \langle C \rangle}{\partial z^2} = \langle R(C) \rangle \qquad [2.39]$$

where

– U is the mean velocity of the fluid;

– E_z is the axial dispersion coefficient.

We thus state:

– $t_S = \dfrac{L}{U}$, the residence time of the fluid in the reactor;

---

4 Average over a reactor's cross-section.

$-Z = \dfrac{z}{L}$, a non-dimensional coordinate equal to 0 at the inlet and 1 at the outlet;

$-\tau = \dfrac{t}{t_S}$, a non-dimensional time equal to 0 at the inlet and 1 at the outlet.

The transport equation [2.39] becomes:

$$\frac{U}{L}\frac{\partial \langle C \rangle}{\partial \tau} + \frac{U}{L}\frac{\partial \langle C \rangle}{\partial Z} - \frac{E_z}{L^2}\frac{\partial^2 \langle C \rangle}{\partial Z^2} = \langle R(C) \rangle \qquad [2.40]$$

Or by multiplying the equation by L/U:

$$\frac{\partial \langle C \rangle}{\partial \tau} + \frac{\partial \langle C \rangle}{\partial Z} - \frac{1}{Pe}\frac{\partial^2 \langle C \rangle}{\partial Z^2} = t_S \langle R(C) \rangle \sqrt{b^2 - 4ac} \qquad [2.41]$$

where $Pe = U.L/E_z$ is the Peclet number that measures the ratio between the characteristic time for dispersion L^2/E_z and for convection L/U.

Let us now place ourselves in a steady state with a large Peclet number, which means that the convection term dominates in relation to the axial dispersion term. It may be observed that the distance z traversed by the fluid from its inlet is given by the product U.t, from which we obtain:

$$\frac{\partial \langle C \rangle}{\partial Z} = t_S \langle R(C) \rangle \Rightarrow \frac{\partial \langle C \rangle}{\partial t} = \langle R(C) \rangle \qquad [2.42]$$

We thus demonstrate the formal equivalence between the equation describing the axial evolution of the concentration in a high Peclet number plug flow reactor and the equation describing the temporal evolution of the closed reactor's concentration. Equation [2.42] can therefore serve as a hydrodynamic model to describe a closed reactor or a plug flow reactor.

An interesting way of considering this equivalence is to isolate a band of the reactor (in dotted lines in Figure 2.18) and to follow it in its displacement at velocity U. As it advances at the fluid's velocity and as we have neglected the axial dispersion term, this band does not exchange matter with the exterior and can be seen as a closed system. This way of thinking in which we follow a fluid volume in its movement is named the *Lagrangian approach*. In contrast, when the observer is interested in a fixed spatial zone, swept by the fluid, we refer to it as an *Eulerian approach*.

2.4.3.2. Micromixing model

The mixing model combines a representation of the concentration distribution and a set of equations to describe its evolution. In this example, it is a question of representing the concentration distribution in a cross-section of the reactor in the form of two Dirac distributions. This proposition amounts to considering that a band of fluid can be subdivided into two zones describing the evolution of the average concentration in this band of fluid. Initially (or in an equivalent way: at the reactor's inlet), one of these two zones contains microorganisms and no substrate, and the other contains substrate and no microorganisms. As time passes, an exchange of matter takes place between these zones under the effect of mixing by diffusion and/or by turbulent transport. The writing of the material balances in each zone – of which the volume is assumed to be constant – leads to the following system formed of four equations:

$$\frac{dX_i}{dt} = \frac{1}{\tau_m}\left(\langle X \rangle - X_i\right) + \mu_{max}\frac{S_i}{k+S_i}X_i \quad i=1,2$$

$$\frac{dS_i}{dt} = \frac{1}{\tau_m}\left(\langle S \rangle - S_i\right) - \frac{1}{Y_{SX}}\mu_{max}\frac{S_i}{k+S_i}X_i \quad i=1,2$$

[2.43]

Index i corresponds to the zone considered; the constant $1/\tau_m$ characterizes the dynamics of mixing between the two zones. The average concentration in cross-section is calculated simply:

$$\langle S \rangle = \frac{S_1 V_1 + S_2 V_2}{V_1 + V_2}$$

$$\langle X \rangle = \frac{X_1 V_1 + X_2 V_2}{V_1 + V_2}$$

[2.44]

This modeling, known as the "Interaction by Exchange with the Mean" IEM model, appeared in the 1970s in the work of Jacques Villermaux [VIL 72]. Later, the same team extended the study by looking at the case of Michaelian reactions [PLA 78].

2.4.3.3. *Analysis*

We can start by observing that the proposed model degenerates into a perfectly mixed reactor model when the mixing time tends towards 0. In fact, since the evolution rate is finite, it is necessary that the difference between the value in each compartment and the mean tends towards zero when the ratio $1/\tau_m$ tends towards infinity. It is an interesting property because it confers on this model a generic character enabling it to cover both the case of the homogeneous reactor and that of the segregated reactor.

In the literature relating to this field, the results are often presented in a non-dimensional form by introducing the following quantities:

$$\eta = \frac{\langle S \rangle_0}{\langle X \rangle_0}, \quad \chi = \frac{\tau_R}{\tau_m}, \quad \theta = \frac{t}{\tau_m}$$

$$\hat{X} = \frac{\langle X \rangle}{\langle X \rangle_0}, \quad \hat{S} = \frac{\langle S \rangle}{\langle S \rangle_0}$$

[2.45]

The curves produced in this form condense so much information that it is quite difficult for the novice to get an idea of the link with the modeling of a real reactor. We propose here an approach that we hope is complementary. First, we can rewrite the model by showing the characteristic times:

$$\frac{dX_i}{d\theta} = \left(\langle X \rangle - X_i \right) + \frac{\tau_m}{\tau_{RX}} X_i, \quad i = 1,2$$

$$\frac{dS_i}{d\theta} = \left(\langle S \rangle - S_i \right) - \frac{\tau_m}{\tau_{RS}} S_i, \quad i = 1,2$$

[2.46]

We know that, in a bioreactor, the growth time is significantly higher than the assimilation time. As Shao Jian Ye notes in his dissertation: *Individual biochemical reactions are usually very fast but overall reaction, e.g. growth, is relatively slow* [YE 85].

To progress, it is now necessary to give a characteristic time-scale for mixing. The model is based on an average description of the mixing at the micro-scale. Consequently, the relevant characteristic time is that of micromixing; it is of the order of 10–100 milliseconds in a turbulent flow. It is therefore possible to conclude that the average concentration of the microorganisms will not change because of growth over the micromixing characteristic time-scale; we will simply see a homogenization of the cell density between the two zones under the effect of the first term of the equation. On the other hand, the assimilation characteristic time can be comparable to that of micromixing, and the second term of the equation for S is no longer negligible. In this case, the rate of consumption in each zone will be impacted by the rate of mixing between the zones. We can only solve the equations for the substrate by considering a constant concentration of microorganisms.

2.4.3.4. Results

The results presented here were obtained with the EDD model, a slightly modified version of the previous model [BAL 90]. The principle remains quite similar except that the volume of the substrate-rich zone increases gradually as it incorporates fluid from the cell-rich zone. The authors of the model proposed a volume evolution as a function of the fluid's turbulent kinetic energy ε:

$$\frac{dV}{dt} = 0.058 \left(\frac{\varepsilon}{v} \right)^{1/2} V \qquad\qquad [2.47]$$

This situation is illustrated in Figure 2.19: the particular conditions at the injection point (fluid flow, local turbulence, pipe diameter) define the initial size, V_0, of the volume of fresh fluid (here the substrate). During transport by the average fluid movement, the turbulent stirring stretches the volume of the fresh fluid; the external fluid is incorporated in the initial volume which passes from V_0 to V_1. The diffusion mechanism helps homogenize this volume between two incorporations. In this figure, we have also illustrated that the cells incorporated in a volume of fresh fluid are suddenly exposed to a large increase in the substrate concentration. The progressive increase in the volume dilutes the substrate, which is also partly consumed. Thus, the next cells incorporated in V_1 will face a variation in the concentration of a lesser magnitude. The fraction of the total population of the reactor exposed to these events is small but the latter are repeated continuously.

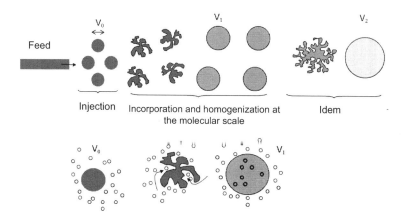

Figure 2.19. *Schematic representation of the mixing mechanism by incorporation. Appearance of a subpopulation of cells exposed to high substrate concentration is shown in bold*

The model presented was used to simulate the experiments carried out by Dunlop and Ye in which yeast circulates between two zones [DUN 90]. One is perfectly stirred, and the other is constituted by a plug flow reactor in which the supply of the substrate takes place. Unexpectedly *a priori*, the addition of a static mixer in the piston area resulted in a very marked biological response symptomatic of an exposure to high substrate concentrations. On the other hand, nothing like this is observed without the static mixer.

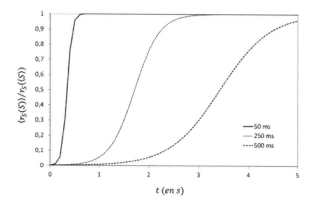

Figure 2.20. *Evolution of the ratio between the substrate's actual assimilation rate calculated using a micromixing model (EED) and that based on a perfect mixing hypothesis (case of a plug flow reactor). For a color version of this figure, see www.iste.co.uk/morchain/bioreactor.zip*

Here, only the piston area has been simulated. The values used are similar to those in Dunlop and Ye's work. Figure 2.20 shows the ratio between the substrate's mean consumption rate calculated by the EDD model and the rate that would be observed using a premixed plug flow reactor hypothesis. The three curves represent different micromixing times. The simulation time corresponds to the residence time of the reactor used in the experiments. These results show that the addition of a static mixer makes it possible to reduce the micromixing time in the zone situated downstream of the substrate injection. Therefore, all cells are exposed to a very high substrate concentration and an intense metabolic response can be observed. On the contrary, in the absence of a static mixer, the micromixing time is substantially longer and does not allow for the intimate mixing of the substrate and the cells in the plug flow reactor. Thus, although the average concentration is high, it does not exist in the reactor; the latter remains highly segregated, so that few cells actually see the substrate injected. When this fluid reaches the strongly stirred zone, the substrate is diluted throughout that zone's volume: despite the addition of concentrated substrate in the plug flow zone, the proportion of cells exposed to this high concentration remains minute.

We have seen that the dynamics of the mixing may depend on the concentration and the flow rate of the feed. We have just shown how the mixing dynamics can affect the apparent rate of biological reactions. Thus, we can imagine that a set of experiments carried out in a chemostat at the same dilution rates generate different segregation states. This dataset can therefore contain a source of non-negligible variability and thus render its use approximate for identifying the parameters of the kinetic laws. In order to compare the results from various biological experiments, it is necessary to check the operating conditions and reactor design very carefully. Besides the feed flow rate, its concentration, and the location of the feeding point as well as the power input are equally important.

2.5. Conclusion

There are many experiments that demonstrate the effect of micromixing on the behavior of microorganisms. The intrinsic nonlinearity of intracellular exchanges and reactions therefore leads to a problem of micromixing. A dual problem of modeling and identification of parameters of the kinetic law follows [MOR 13]. These two aspects are illustrated in Figure 2.21. In this case, the mean concentration is identical at all points but the local concentration distribution varies in space. If we take into account this local distribution to calculate the reaction term, it will also be variable in space. On the other hand, the modeling of the reaction term is based only on the mean concentration in each cell and the micromixing effect is not represented. It is this approach that is generally retained when coupling the computational fluid mechanics using a RANS approach with a biological model (whichever it is). Moreover, close to the injection site, segregation is strong, the distribution of concentration has the form of a "U" (see [FOX 03]) and the error based on the average concentration is then very large [LIN 12b].

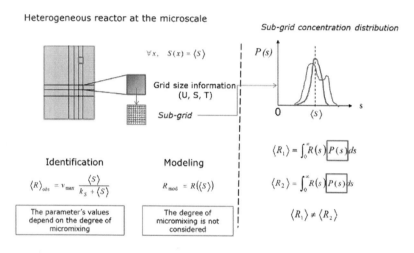

Figure 2.21. *Consequences of poor micromixing on modeling and identifiability of parameters of biological kinetics. For a color version of this figure, see www.iste.co.uk/morchain/bioreactor.zip*

From the experimental point of view, only the average concentration in the reactor is generally accessible. Bylund *et al.* [BYL 98] implemented a local measurement in an 8 m^3 industrial bioreactor. However, this is an average value on the scale of the measurement volume, which is very large in relation to the cells. The local distribution seen by the cells thus remains in general inaccessible. However, if such a distribution exists, the apparent rate of the biological reaction is impacted. Thus, the micromixing state impacts the identification of the parameters of the kinetic law used, namely the maximum assimilation rate and the affinity for the substrate.

It should also be noted that since the local concentration distribution results from the competition between the mixing and assimilation rates, the state of micromixing necessarily evolves towards greater segregation when the cell concentration increases. However, the purely kinetic vision is insufficient. Indeed, different intracellular reactions will take place according to the intensity of the mass fluxes received by the cell. Thus, the rate and also the nature of the reactions are variable as a function of the mixing state [LIN 14].

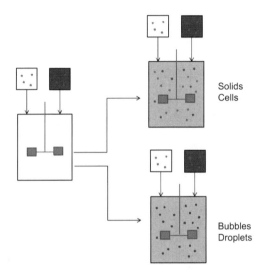

Figure 2.22. *Effect of the nature of the dispersed phase on the representation of mixing in a multi-phase system. For a color version of this figure, see www.iste.co.uk/morchain/bioreactor.zip*

To conclude, let us return to an essential aspect of bioreactors: their inherently heterogeneous and multi-phase character. Let us examine this notion from the angle of mixing two fluid currents, each containing two phases at different concentrations as illustrated in Figure 2.22. The question of mixing arises for both phases: the liquid phase and the dispersed phase composed of bubbles/drops or solid particles/cells. In a highly stirred reactor, it can be assumed that, in the first case, the two phases will be rapidly homogenized because the bubbles/drops are subjected to rupture and coalescence. On the other hand, if the dispersed phase consists of solid particles or cells, the intimate mixing of the particles/cells between themselves is not carried out. Each particle retains its own identity. A new notion appears: the relaxation time of the biological particles, that is to say the time necessary for the cells to reach a state of equilibrium with their new environment resulting from the mixing of the liquid phases.

Thus far, we have assumed that the biological system is instantaneously in equilibrium and there is no effect on how the microorganisms experienced the concentration distribution. In other words, microorganisms were supposed to adapt immediately to their environment. There is clearly a very strong link between the reactivity of a biological system and the variations in concentrations undergone in the reactor. These variations result from both the concentration volume distribution and the velocity field which transports the particles. It is therefore possible to characterize the heterogeneity of the biological reactor using the ratio between the maximum concentration deviation divided by the macromixing time $\frac{(\Delta C)_{max}}{\tau}$.

By adopting a Lagrangian view this time, this quantity can perfectly be interpreted as the rate of concentration change along the trajectory followed by a microorganism.

$$\frac{\left(\Delta C\right)_{max}}{\tau_C} \equiv \nabla C_{@p} \frac{\partial \overrightarrow{x_p}}{\partial t} \qquad . \qquad\qquad [2.48]$$

The appearance of a consequence in metabolic terms will therefore be linked to the ability of microorganisms to respond sufficiently quickly to the change imposed by the existence of a velocity field and a concentration field. In the case of a rapid response, the biological system can be considered to be in equilibrium with its environment. It is not necessary to distinguish individuals according to their history. In the opposite case, disequilibrium appears and can induce consequences at the metabolic level, it is necessary to keep track of each group of individuals and its particular state. We will see in the next chapter which biological phenomena can play a role in response to these concentration fluctuations.

3

Assimilation, Transfer, Equilibrium

3.1. Introduction

Since the origin of the multi-phase point of view, the notion of biological reaction has meaning only within the biological phase. The key elements become, on the one hand, the description of the intracellular reactions and, on the other hand, the determination of the exchanged fluxes between cells and other phases, known as liquid–cell transfer or assimilation. In this chapter, we will review the literature that allows us to identify a specificity of living systems: the ability of living organisms to modulate their capacity for assimilation, that is, to regulate transfers through the membrane. This ability to modulate transfers is a difficult point to model. Given the diversity of biological adaptation mechanisms in nature, we will focus on some studies by identifying what may constitute a generic property for microorganisms and examining the consequences from the modeling viewpoint.

It will then be interesting to examine the notion of balanced growth and the ability of biological models to describe an unbalanced situation. We can already state that the proper function of a living system is, among others, to be able to maintain itself outside the state of equilibrium, that is to say, to devote energy to control exchanges with the outside world. In this respect, the biological phase differs fundamentally from the other phases, for which laws governing thermodynamic equilibrium at the interfaces are available.

In most of the common models, the assimilation rate is deduced from the growth rate calculated on the basis of the Michaelis–Menten-type enzymatic kinetics [SCH 03, LAP 06]. It is therefore deduced from the liquid-phase concentrations. Unfortunately, it appears that the assimilation law identified in a well-mixed reactor must be modified to describe what happens in a large, more heterogeneous reactor [VRÀ 01]. This clearly indicates that some aspects of the mass transfer between the liquid phase and cells are not adequately addressed in the models. The lack of robustness of bioreactor models is largely due to the incorrect estimation of assimilated fluxes.

In fact, multiple transport systems act in parallel. Their activity is maximum at different concentrations. The overlapping of their range of activity allows the cell to simultaneously use a combination of carriers. A new complexity thus arises because the contribution of each system to the assimilated flux evolves over time in response to changes in the concentration in the cells' environment. Thus, the concentration distribution affects the assimilation capacity of different individuals of the same population.

3.2. Transfers between phases

3.2.1. Gas–liquid transfer

First, we will consider the case of the transfer of a gaseous compound (oxygen) to the liquid phase, followed by its assimilation by the biological phase. For this purpose, we consider a fluid volume characterized by a volume fraction of gas (ε_G) and a volume fraction of cells (ε_S). The equation of conservation of the oxygen mass in the liquid phase (see Chapter 1, equation [1.11]) is as follows:

$$\frac{\partial \varepsilon_L O_2 V}{\partial t} = F_{L,i} O_{2,i} - F_{L,o} O_{2,o} + r_{O_2}\left(\mathbf{C}_L\right) V_L + \varphi_{GL} + \varphi_{SL} \tag{3.1}$$

$$\varepsilon_L + \varepsilon_G + \varepsilon_S = 1$$

For a closed system, the flux exchanges with the exterior are zero. Moreover, in the absence of a chemical reaction in the liquid phase and following the developments made in Chapter 1, we obtain an equation of the form:

$$\frac{\partial O_{2L}}{\partial t} = K_{GL} \frac{a}{\varepsilon_L}\left(O_{2L}^* - O_{2L}\right) - q_{O_2} \frac{1-\varepsilon_G}{\varepsilon_L} X \tag{3.2}$$

For simplicity, we set aside the volume fractions, while understanding that the liquid fraction is close to 1 and the gas fraction is of the order of a few percent. We can now examine the solution of this steady-state equation.

The oxygen flux transferred from the gas to the liquid is counterbalanced by a flux of assimilated oxygen or transferred from the liquid to the cells:

$$K_{GL}a\left(O_{2L}^{*}-O_{2L}\right)=q_{O_2}X \tag{3.3}$$

Before proceeding further, we recall that the specific assimilation rate (q_{O2}) is treated here as a property of the cell. Its value can be modulated by the cell according to its needs. In equilibrium (from the viewpoint of the cell), the assimilation rate makes it possible to satisfy the need exactly. The question related to the assessment of this need is being put aside for now. The equality between transferred and assimilated flow covers the following three very distinct situations:

– The concentration of oxygen in the liquid is zero. The oxygen flux transferred to the cells is $\varphi_{O_2}=K_{GL}aO_2^{*}$. This situation is further subdivided into two sub-cases:

- The oxygen flux transferred by the gas exactly balances the need of the cells. We cannot speak of limitation in the strict sense insofar as the needs of the microorganisms are satisfied: $\varphi_{O_2}=q_{O_2}X$.

- The transferred oxygen flux is insufficient $\varphi_{O_2}<q_{O_2}X$. The respiration is effectively limited by the gas–liquid transfer, and the oxygen need is not satisfied. The specific rate of effective assimilation is then given by the equation:

$$q_{O_2}{}^{eff}=\frac{K_{GL}aO_{2L}^{*}}{X} \tag{3.4}$$

It is important to note the similarity with the expression established for a reaction rate limited by the mixing, as introduced in Chapter 2. Here, again the intensity of the limitation is related to the quantity of cells present in the medium.

– The oxygen concentration in the liquid is non-zero. In this case, a correlation can be established between the specific assimilation rate in the steady state[1] and the concentration of dissolved oxygen in the liquid phase. The results are then formulated using a law similar to that used for other substrates:

---

1 This precision is crucial.

$$\overline{q_{O_2}} = \overline{q_{O_2,max}} \frac{\overline{O_{2L}}}{k_{O2} + \overline{O_{2L}}} \qquad [3.5]$$

The notation $\overline{}$ indicates that these are quantities in the steady state. This is a relationship at equilibrium, a thermodynamic equilibrium law in some way. Then, the liquid-phase concentration can be calculated as the equation's solution [3.6]:

$$K_{GL}a\left(O_{2L}^* - \overline{O_{2L}}\right) = \overline{q_{O_2,max}} \frac{\overline{O_{2L}}}{k_{O2} + \overline{O_{2L}}} X \qquad [3.6]$$

It shall be noted that the above expression only makes sense if the oxygen flux required by the cells is less than the maximum flux transferable by the gas. This will result in a non-zero equilibrium concentration in the liquid.

If we adopt expression [3.5] to express the transient oxygen consumption, it is assumed that the cells' need decreases as fast as the oxygen concentration dissolved in the liquid. The asymptotic behavior of this expression generates a discontinuity in the neighborhood of zero. In other words, it is possible to calculate the equilibrium concentration by solving equation [3.6], but obtaining the numerical solution of the dynamic equation [3.7] will be more difficult as the final value approaches zero. Indeed, the assimilated flux approaches zero when the concentration of oxygen approaches zero, while the gas–liquid transfer flux tends to its maximum value:

$$\frac{\partial O_{2L}}{\partial t} = K_{GL}a\left(O_{2L}^* - O_{2L}\right) - q_{O_2,max} \frac{O_{2L}}{k_{O2} + O_{2L}} X \qquad [3.7]$$

One possibility is to replace equation [3.7] with the algebraic equation [3.4] when the oxygen concentration tends to zero.

The numerical integration of equation [3.7] poses a serious problem when the oxygen concentration approaches zero. Indeed, if the concentration calculated at time t approaches zero, the term describing the assimilation decreases, whereas the transfer term causes a sharp rise in the oxygen concentration. The greater the oxygen transfer coefficient, the higher the increase in the oxygen concentration. Following the increase in the oxygen concentration, assimilation resumes and causes a sudden drop in concentration during the next time step. This causes numerical oscillations in the oxygen concentration value. As the concentrations cannot be negative, the temporal integration of the equation is marked by strong numerical instabilities and proves to be very costly in terms of computation time, because it is necessary to use very short time steps, adapted to the dynamics of the fastest phenomenon (most of the time, it will be the transfer term). A numerical trick is to reset to zero the concentrations that would have become negative at the end of a time step, but this is only a tool used to overcome modeling defects.

It is thus observed that the calculation of the assimilated flux must take into account the oxygen concentration if it is non-zero, but it must be based on the gas–liquid transfer flux if the latter constitutes the limiting factor. However, the classical models are virtually all formulated in terms of concentrations. Numerical difficulties stem from the fact that transfer physics is not correctly modeled. More precisely, assimilation is described in a kinetic way, whereas the underlying key phenomenon is in fact the transfer from the liquid to the cell.

In order to circumvent this difficulty, the following formulation can be adopted: the assimilated specific flux corresponds to the minimum between the flux associated with the cell's respiratory capacity and the maximum flux related to the gas–liquid transfer:

$$\varphi_{O2} = \min\left(q_{O_2}, \frac{K_L a O_{2L}^*}{X} \right) X \qquad [3.8]$$

This expression is not continuous and involves the cells' specific assimilation rate (q_{O_2}), which remains to be defined. There are multiple ways to estimate this value:

– Consider it as a constant;

– Link it algebraically to the growth rate $q_{O_2} = Y_{OX}\mu$;

– Treat it as a property of the cells and write a law of evolution for this new property. If we know the value of the assimilation rate in the steady state, that is to say, when the cell has adapted its capacity to its needs and to the availability of oxygen in the medium, we can write a very simple relaxation equation toward the equilibrium value:

$$\frac{dq_{O_2}}{dt} = \frac{1}{\tau}\left(\overline{q_{O_2}} - q_{O_2} \right) \qquad [3.9]$$

By plugging equation [3.8] into equation [3.2] for the consumption term, we get a new balance equation for oxygen in liquid phase, supplemented by a closure law (e.g. [3.9]), making it possible to calculate the cells' specific assimilation rate:

$$\frac{\partial O_{2L}}{\partial t} = K_{GL}a\left(O_{2L}^* - O_{2L}\right) - \min\left(q_{O_2}, \frac{K_L a O_{2L}^*}{X}\right)X$$

$$\frac{dq_{O_2}}{dt} = \frac{1}{\tau}\left(q_{O_2} - \overline{q_{O_2}}\right)$$

[3.10]

It will be observed that in the steady state, the second equation leads to equality $q_{O2} = \overline{q_{O2}}$.

Moreover, if the concentration is non-zero, we necessarily have $\overline{q_{O2}} < K_{GL}aO_{2L}^*/X$, hence equation [3.11] whose solution will provide the value of the dissolved oxygen concentration:

$$K_{GL}\frac{a}{\varepsilon_L}\left(O_{2L}^* - \overline{O_{2L}}\right) = \overline{q_{O_2}}X$$

[3.11]

or

$$\overline{O_{2L}} = O_{2L}^* - \frac{\varepsilon_L}{K_{GL}a}\overline{q_{O_2}}X$$

[3.12]

If the concentration of the dissolved oxygen is zero, we are in an assimilation regime that is limited by the transfer and equation [3.4] is derived.

3.2.2. Liquid–microorganisms transfer

The transport of solutes through the cell membrane obeys various mechanisms illustrated in Figure 3.1 taken from Bailey and Ollis [BAI 86].

3.2.2.1. Simple diffusion

This is a physical phenomenon in the classical sense; the transport of the solute takes place in the direction opposite to the gradient (from high to low concentrations) through the membrane (consisting mainly of lipids). The diffusion flux density $(\overrightarrow{J_D})$ is therefore in practice dependent on the diffusivity of the solutes in the lipids, D (in $m^2.s^{-1}$) and is proportional to the concentration gradient on either side of the membrane of thickness l:

$$\overrightarrow{J_D} = \frac{D}{l}(C_{ext} - C_{int})\vec{e} \qquad [3.13]$$

where $\vec{e}$ is the unit vector in the frame bound to the cell. Thus, the absorption flux corresponds to the integral of this flux density on the cell's surface, of measurement A:

$$q = \iint \overrightarrow{J_D}.\vec{e}da = \frac{D}{l}(C_{ext} - C_{int})A \qquad [3.14]$$

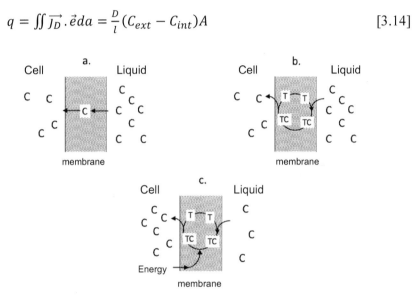

Figure 3.1. *Types of membrane transport: a) diffusion; b) facilitated diffusion; c) active transport, by Bailey and Ollis [BAI 86]*

3.2.2.2. Facilitated transport

Some charged molecules, which are rarely soluble in lipids, nevertheless transfer at high speeds. This is explained by the presence of molecules called "carriers" that associate themselves specifically with the solute, transporting it in the membrane and release it inside the cell. The overall transport rate is this time limited by the formation and dissociation kinetics of the carrier-solute complex and by the transport of the complex into the membrane. From the kinetic viewpoint, the reaction would be as follows:

$$C_{ext} + T \xleftrightarrow{\quad} TC$$
$$TC \longrightarrow C_{int} + T$$

[3.15]

The usual pattern of an enzymatic reaction is thus recognized, from which it can be concluded that the transport rate should obey the Michaelis-Menten-type law, with carrier protein T playing the same role as an enzyme:

$$q = v_{max}(T)\frac{C_{ext}}{K_S + C_{ext}}$$

[3.16]

3.2.2.3. Active transport

Active transport is distinguished from facilitated transport by two aspects: it is carried out opposite to the concentration gradient (from the least concentrated to the most concentrated), and requires energy expenditure. However, the usual pattern of an enzymatic reaction remains and the transport rate still obeys the Michaelis–Menten-type law:

$$q_{TA} = v_{max}(T)\frac{S_{out}}{K_S^* + S_{out}}$$

[3.17]

This time, the constant K_S^* integrates the fact that reactions generating energy are coupled to the transport process. We shall see in the next paragraph that this may lead to a more complex equation of the velocity law.

NOTE.– Regarding the Michaelis–Menten law, Bailey and Ollis reported that the maximum velocity depends in fact on the total enzyme concentration (free and complex) denoted e_0.

Applied to the case of carriers, this means that the maximum assimilation rate will be correlated with the carrier concentration (or the number of carriers per cell). We will later see that the activity of these systems is related to the concentration seen by the cell.

Moreover, the velocity law assumes that the quantity of the substrate largely exceeds that of the enzyme, which is false when we have very low substrate concentrations (unlikely in an industrial bioreactor).

Active transport effectively imposes a coupling between mass and energy conservation equations at the cell scale. A cell cannot sustainably spend more energy on assimilating a substrate than it retrieves by transforming it.

3.2.2.4. Carriers operating in parallel

The results of Ferenci [FER 96, FER 99a, FER 99b, FER 07] on the assimilation mechanisms of carbonaceous substrates (mainly in *Escherichia coli*) are reproduced here. Three groups of mechanisms responsible for substrate transport have been identified in the case of Gram-negative bacteria:

– The so-called high-affinity system (efficient at low concentrations) drawing its energy from the hydrolysis of ATP and involving a binding protein located in the periplasm, called permease. It is an active-transport-type system.

– A phospho transferase system (PTS) with intermediate affinity for the substrate. This system is controlled by the ratio of two compounds (phosphoenolpyruvate/pyruvate). Two types of modeling are proposed for this system:

- A Michaelis–Menten-type law:

$$q_S^{PTS} = q_{S,\max}^{PTS} \frac{S}{K_S + S} \qquad [3.18]$$

- A more precise formulation explicitly involves the concentration ratio of two intracellular compounds: phosphoenolpyruvate/pyruvate [CHA 02]. Here, we are dealing with a structured modeling; see section 3.4.2:

$$q_S^{PTS} = q_{S,\max}^{PTS} \frac{S \dfrac{C_{pep}}{C_{pyr}}}{\left(K_1 + K_2 \dfrac{C_{pep}}{C_{pyr}} + K_3 S + S \dfrac{C_{pep}}{C_{pyr}} \right)\left(1 + \dfrac{C_{g6p}}{K_4} \right)} \qquad [3.19]$$

– A low-affinity system (effective at high concentrations), called a *symporter*, in which the transport of the substrate is coupled with an ion exchange through the membrane.

An illustration of these mechanisms is presented below. It should be noted that sugar transport systems are not all specific; the cell thus acquires the means to assimilate a multitude of carbon compounds. It also appears that these systems are likely to work in parallel.

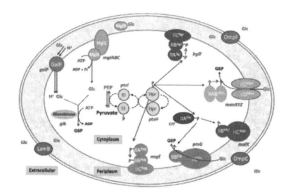

Figure 3.2. *Schematic representation of the different systems of transfer of sugars through the membrane of an E. coli bacterium, taken from the works of Fuentes et al. [FUE 13]. For a color version of this figure, see www.iste.co.uk/morchain/bioreactor.zip*

Because these different transport mechanisms can be simultaneously active, the assimilation capacity[2] is generally written as the sum of the specific assimilation rates that can be achieved by each system:

$$q_S = q_S^{PTS} + q_S^{por} + q_S^{sym} \qquad [3.20]$$

3.2.2.5. *Carriers capable of modulating their activity*

The activities of the various systems responsible for the transport of glucose through the cell membrane depend on the concentration of glucose in the medium (liquid phase). Each system is more particularly adapted to a concentration range, which overlaps in order to ensure continuity in the overall assimilation capacity [FER 99b]. We thus distinguish between non-limiting cultures conditions with a high sugar concentration, limiting conditions in which the cells are undernourished (hungry bacteria), and the so-called starvation conditions that are highly stressful for microorganisms [FER 01]. In each case, the set of carriers are active, but their contribution to the observed overall assimilation rate is different. Most of the assimilation is ensured by the system best suited to the particular conditions.

---

2 Whether this capacity is fully exploited or not is still to be determined.

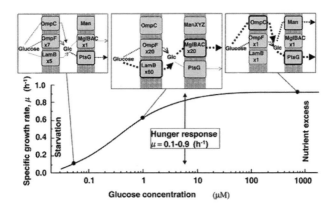

Figure 3.3. *Hunger resulting in a modulation of membrane transport activity in E. coli; figure taken from Ferenci [FER 01]*

Figure 3.3, taken from Ferenci [FER 01], illustrates the fact that the different carriers are active simultaneously, but participate at varying levels in the assimilated total flux depending on the available concentration in the liquid medium. The three diagrams above the curve show the components of the membrane affected by the hunger response, specifically OmpF, OmpC and LamBporins. Through the cytoplasmic membrane, the main adaptation concerns the Mgl-dependent transport system. The levels of transcription of the genes associated with Mgl, LamB and OmpF in the three nutritional states corresponding to different growth rates are also illustrated in the figure. The arbitrary value 1 is assigned to excess substrate conditions. The other values relate to this reference state. It is thus noted that the strongest response in terms of gene expression is for intermediate growth rates around $\mu = 0.6 \; h^{-1}$. Finally, the thickness of the line is relative to the contribution of each system, which reveals the flux changes in the different paths of glucose transport as a function of the growth rate. These data were acquired during chemostat cultures.

In addition, the work of T. Egli on prolonged cultures in chemostat with very low substrate concentration will be discussed with interest. This work shows that, in addition to the dynamic adjustment of carriers, mutations produce groups of individuals better suited to starvation conditions, so that the assimilation capacities of the population improve overall during long time periods [KOV 98, SEN 94]. Thus, as T. Ferenci pointed out, the steady-state concept, which is inherently attached to the chemostat, is at best illusory and in fact a constant decrease in the residual concentration is observed when the cultures are prolonged over several hundred generations. Under prolonged conditions of undernourishment, populations become highly heterogeneous and their genotype is significantly different from that of the original population of the culture [FER 01].

3.2.3. Synthesis and conclusion

In the remarkable work of Natarajam and Srienc [NAT 00] on the assimilation of a non-metabolizable glucose substitute, the following conclusion emerges:

> *In a continuous culture (dilution ratio between 0.1 and 0.4 h^{-1}), the capacity of glucose assimilation is independent of the dilution rate, while the glucose concentration in the medium decreases with the dilution rate. Thus, the intrinsic assimilation capacity remains unchanged in this range of dilution rates and the rate of assimilation is limited by the availability of glucose. While cells that develop slowly possess the same ability to assimilate glucose (if provided) as cells growing faster, it is possible that they will not be able to use this sugar as effectively, because their physiological state [...] is different.*

Thus, cells that develop slowly in a medium lacking glucose show an identical ability to assimilate glucose to those cells developing more rapidly in a rich environment. However, this assimilation capacity is not fully exploited because the concentration in the medium is low. Thus, the uptake rate actually decreases with the concentration in the medium; however, this is not due to a biological limitation and we can think that assimilation is limited by micromixing.

To summarize, the activity of high-affinity carriers (effective at low concentrations) increases as the concentration in the medium decreases. This passage between the different modes of transport allows the cell to maintain a constant assimilated flux despite the decrease in the external concentration. Of course, this adaptation is never instantaneous and a sudden decrease in concentration does not entail the same consequences as a slower reduction solely induced by the cells' consumption. In the reverse situation – a sudden increase in the concentration applied to cells accustomed to a substrate-depleted environment – the cells will fully exploit their high assimilation capacity without being able to transform this surplus of sugar into metabolites necessary for growth. In such a case, the uptake and growth rate are temporarily de-coupled.

Transport systems therefore adapt dynamically to environmental conditions and to the growth rate of the cells. This adaptation involves various regulation mechanisms having different characteristic times. Repeated exposure to nutritionally limiting conditions tends to produce heterogeneity within microbial populations.

3.3. Equilibrium or dynamic responses: experimental illustrations

The first point to be raised concerns the multitude of phenomena taking place within the cell in response to a controlled or non-controlled perturbation. The associated characteristic time range is very broad and the level of observability evolves in parallel. As with turbulence, we can associate a biological response with a size and time scale. Thus, variations in the growth rate can be assessed at the reactor scale and over several hours; by contrast, some metabolic responses can be very rapid, localized within the cell itself and require the dosing of intracellular metabolites in trace quantities [MAR 12]. The notion of biological response is therefore closely linked to the experimental protocol and the type of measurement (variable and equipment).

The second point concerns the couplings within the cell: the response to a perturbation observed via a measurement (usually an extra or intracellular dosage) is rarely that of a single and isolated biological phenomenon. The measurement most often reflects the consequences of a more or less extensive set of coupled sub-systems responding in cascade. The behavior is highly nonlinear: the activation and deactivation of a biological process respond to different time constants, called induction time and relaxation time, respectively.

Third, the responses observed in a bioreactor with concentration gradients are those of a population that is repeatedly exposed to variable frequency and amplitude perturbations. Experiments consisting of imposing a step-up (of concentration and/or dilution rate) or concentration pulse [GUI 04, SUN 12] are complementary to those carried out in multi-stage reactors [BEN 06, HEI 15, NEU 95].

Finally, even in the relatively simpler case of a homogeneous laboratory reactor, we have seen that the culture's history, the trajectory before the application of a perturbation, is important, because it determines the state of the population at the moment it undergoes the perturbation.

3.3.1. *Dynamic response in terms of growth rate*

3.3.1.1. *Evolution of the growth rate in response to a sudden increase in the feed rate of an open continuous bioreactor (CSTR)*

We first examine the data of Käterrer *et al.* [KÄT 86] on the sudden change in flow rate. The results are presented in Figures 3.4 and 3.5.

The first striking result is that the nature, amplitude and duration of the response are significantly different depending on the extent of the imposed perturbation. Figure 3.4 shows that substrate and cell concentrations vary very slightly and that the transitory phase does not exceed 4 h after the perturbation. In general, this is what we observed experimentally if the change is of weak amplitude [SHO 03]. The same authors indicate that oscillations around the new steady state can continue for several hours. Such oscillations are also visible here. These variations could, however, be attributed to uncertainty in the measurement of cell concentration.

In Figure 3.5, a response of high amplitude and of a different nature is observed: although the cell concentration had increased in the first experiment, it dropped sharply following a greater increase in the dilution rate. At the same time, the decrease in the cell concentration, and therefore in the substrate consumption, combined with the increase in the flow rate, caused a considerable accumulation of the substrate in the medium. It is clear that the amplitude of variation in flow (and concentration) affects the very nature of the biological response.

In both cases, the final microorganism concentration is different from the initial concentration, which is generally attributed to a variable substrate conversion yield with the specific growth rate (Pirt's law).

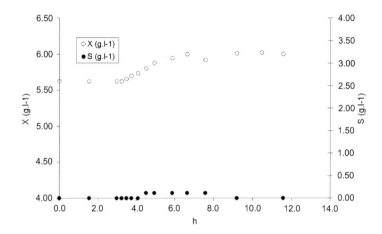

Figure 3.4. *Response of a bacterial culture of Candida tropicalis to a sudden increase in the dilution rate ($Q_L/V = 0.1 \rightarrow 0.3\ h^{-1}$)*

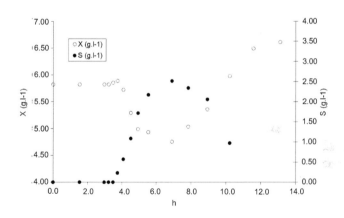

Figure 3.5. *Response of a bacterial culture of Candida tropicalis to a sudden increase in the dilution rate ($Q_L/V = 0.1 \rightarrow 0.42\ h^{-1}$)*

However, it can be seen in the experiments that:

– Following a strong increase in flow rate, the cell concentration decreases significantly. Specific growth rates of cells are therefore lower than the dilution rate in the period $t \in [3.5–7\ h]$. At the same time, sugar accumulates in the reactor, while remaining below a value capable of inhibiting growth. If the algebraic relationship $\mu = f(S)$ is applied, the specific growth rate should be maximum, which is not the case.

– Later, the cell concentration increases again; the specific growth rate is greater than the dilution rate ($t \in [7\text{–}12 \text{ h}]$). The concentration of sugar decreases during this period.

An interpretation from the viewpoint of linear systems shows that the oscillating responses observed are characteristic of a linear system of order equal to or greater than 2. For a low amplitude scale, the response is similar to that of a system of order 1 or 2 having a damping factor close to 1 if the weak oscillations observed around the new steady state are considered significant. For a greater variation of the flow, the nonlinearity of the biological system manifests itself with force. It should be noted that the evolution of the biomass concentration follows a very different trajectory (decrease and then increase) (also visible in Guillou *et al.* [GUI 04]). Other similar studies provide similar conclusions [ABU 89]; the response of a chemostat culture to a sudden change in the dilution rate depends on the initial value of the dilution rate, amplitude and variation sign.

CONCLUSIONS.– Under dynamic conditions, the growth rate of microorganisms does not match the dilution rate.

In a dynamic regime, the specific growth rate is de-coupled from the concentration of substrate present in the medium. The algebraic relationship can no longer be used to calculate the population growth rate.

These conclusions have far-reaching consequences for modeling: the equilibrium hypothesis is admissible for low perturbations. Indeed, because of a small variation in the flow rate, the cell concentration varies very little, which supports the idea that the growth rate has adapted almost instantaneously to the new dilution rate. On the contrary, this assumption is not appropriate if the population is exposed to excessive perturbation. Moreover, this notion of strong perturbation is relative to the population's state (say to its growth rate before the perturbation to simplify). The algebraic relationship between growth rate and concentration is thus an excessive simplification of the functioning of the biological system and fails to describe certain dynamics because it postulates an immediate adaptation of the growth rate of individuals to environmental conditions [MOR 09, SIL 08]. Although it has been denounced on many occasions, this type of relationship interests us by its ease of use. However, it masks the very essence of biological systems: the ability to maintain a system out of equilibrium or in other words the existence of a relaxation time toward the state of equilibrium.

In this field, Perret [PER 60] gave us a remarkable theoretical work, far ahead of their time, which warned against the misuse of Monod's law [MON 49] for the description of the dynamics of biological systems. Perret showed that the inertia of a biological system leads necessarily to the phenomenon known as growth rate hysteresis, that is, the de-coupling between concentration and growth rate in transitory regime. Thus, Perret deduced that in response to a variation in substrate concentration, the population's growth rate does not follow the equilibrium law (solid line in Figure 3.6) but first evolves under this limit before crossing it and going above it when substrate concentration decreases. In this figure, the arrows indicate the course of time.

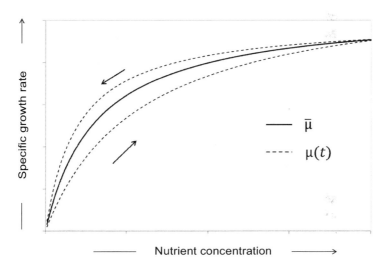

Figure 3.6. *Growth rate hysteresis: curves showing the general shape of the relationship between nutrient concentration and growth rate in an expanding auto-catalytic system. Continuous line: growth rate in limitation by the substrate (intake and consumption equilibrate), dotted line: instantaneous value of the growth rate during an increase and a decrease in the concentration of nutrients. According to Perret [PER 60]*

From these experimental observations and their analysis, it appears that Monod's law, $\mu = f\ (C_L)$, and Pirt's law, $Y = f\ (\mu)$, are similar to thermodynamic equilibrium laws. They provide the relationships between reaction rates and liquid-phase concentrations when microorganisms are in equilibrium with their environment.

Adopting these relationships to describe the dynamics of a bioreactor is equivalent to positing that the trajectory followed to move from one equilibrium point to another is a succession of equilibrium states.

Moreover, this notion of equilibrium is in fact rather fragile. It is attached to an ensemble property (average reaction rate, growth rate). This equilibrium is that of a dynamic system, that is, one of the stable states of a system of coupled differential equations [CHU 96]. This dynamic system has its own characteristics; in particular, it is constituted by numerous coupled sub-systems, each having its own characteristic time, so that it is practically impossible for this system to be constantly in equilibrium with its environment. In other words, and this is probably the most interesting point, it can be observed that a biological system by nature has the ability not to be in equilibrium with its environment, which distinguishes it from a chemical system.

3.3.1.2. *Evolution of the growth rate in response to a progressive increase in the feed rate of an open bioreactor operating in continuous mode (accelerostat)*

The previously cited work indicates that the rate of population growth adjusts to the dilution rate with a finite rate. Adaptation is never immediate. Rather than suddenly altering the dilution rate (step-up of the flow rates), researchers had the idea of increasing it gradually. This type of continuous culture at increasing feed rate is called accelerostat (sometimes called A-stat). It tests the ability of a population to adapt to the speed of change imposed on it.

The idea that the results obtained in chemostat at different dilution rates depend strongly on the rate of change from one dilution rate to another is present in the work of van Dijken and Sheffers [VAN 86] citing Barford and Hall [BAR 79]:

Recent studies have made clear that above a certain dilution rate the transition from respiratory to fermentative catabolism may not be accompanied by decreased rates of oxygen consumption through repression of respiratory enzymes. In contrast to the response seen after a sudden single-step increase in dilution rate, careful stepwise increase may result in establishing steady-state cultures up to μ_{max} which perform alcoholic fermentation at constant, unaffected rate of oxygen consumption. Thus, via careful adaptation, a situation may be reached where the specific rate of oxygen consumption is not decreased but remains constant and independent of the dilution rate.

Figure 3.7 is taken from the work of Adamberg *et al.* [ADA 09] on *Lactococcus lactis* cultures in accelerostat. Several experiments are carried out with increasing values of parameter *a*, which defines the speed at which the flow rate increases. The results indicate that there is a critical value of parameter *a* beyond which the population is no longer able to keep up in terms of specific growth rate. When the flow rate increases too fast, the population no longer has time to carry out the intracellular adjustments, allowing it to increase its growth rate at the same rate as that imposed by the increase in the dilution rate. These experiments therefore provide a means of estimating the maximum rate of adaptation of a population's growth rate.

The upper graph of Figure 3.7 reveals a situation where the hypothesis of quasi-stationary equilibrium between dilution rate and growth rate is verified.

The lower graph in Figure 3.7 illustrates a different situation (greater acceleration). Two successive acceleration phases are also imposed on the population. The first one begins around t = 60 h and lasts 20 h. The growth rate increases first and then decreases. It is noted that the increase in the glucose concentration in parallel with the dilution rate does not lead to an increase in the growth rate in similar proportions. The second acceleration phase is proposed at t = 98 h, whereas the population's growth rate has barely reached the constant dilution rate maintained since t = 86 h. During this second acceleration (the value of *a* is the same), the decline is immediate. It is therefore clear that the state of the population largely determines its response capacity. Characterizing the state of a population through the growth rate alone is therefore insufficient.

We will see later in this chapter how to integrate the de-coupling between concentration and specific growth rate in order to account for this reality.

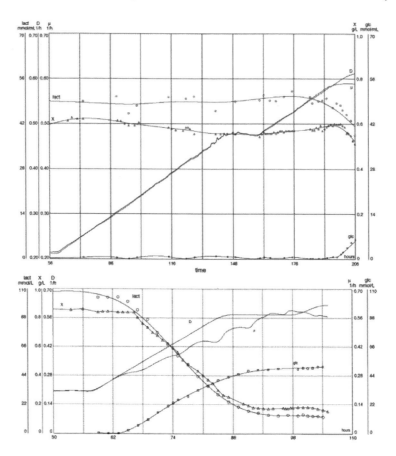

Figure 3.7. *Lactococcus lactis culture IL1403 in accelerostat. Upper graph: a = 0.003 h^{-2} and S$_0$ = 5 g.l^{-1}; lower graph: a = 0.015 h^{-2} and S$_0$ = 10 g.l^{-1} [ADA 09]*

3.3.2. *Dynamic response for the assimilation*

3.3.2.1. *Experimental examples*

We start by presenting results established by Li on *Pseudomonas perfectomarina* cultures [LI 82]. Samples are taken from continuous cultures at different dilution rates and put into contact with carbon-14-

labeled bicarbonate. The accumulation of radioactivity in the cells makes it possible to evaluate the carbon's assimilation rate. In parallel, the evolution of the optical density of the samples makes it possible to follow the growth. Three distinct trends can be seen in Figure 3.8. The slope of the assimilation curve decreases, remains constant or increases with time. This clearly proves that the assimilation rate evolves over time and differs depending on the growth rate of the population before exposure to a non-limiting amount of carbon. Although the data are presented on two scales, it appears clearly in this figure that the initial slope is higher when the initial growth rate is low.

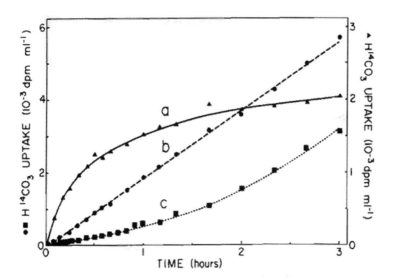

Figure 3.8. *Evolution of assimilation of HCO_3. Effect of population growth rate before perturbation. (a): $\mu = 0.074\ h^{-1}$, (b): $\mu = 0.135\ h^{-1}$, (c): $\mu = 0.181\ h^{-1}$*

These results suggest that in chemostat the assimilation capacity at low dilution rate is high and remains under-exploited because of the low substrate concentration in the culture[3]. To be convinced of this not so intuitive result, it is preferable to examine the dimensionless quantity R defined as the ratio of the specific assimilation rate (here, ions HCO_3^-) to the specific growth rate:

---

3 In chemostat, the residual concentration decreases with growth rate.

$$R = \frac{q_{HCO_3}}{\mu}$$

[3.21]

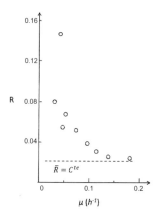

Figure 3.9. *Relationship between the initial assimilation rate (over 15 min) and the specific growth rate as a function of the growth rate before perturbation (according to Li [LI 82])*

The results of Figure 3.9 relate to the specific rate of instantaneous assimilation following the addition of a dose of labeled bicarbonate. It is observed that the instantaneous assimilation rate of carbon is not directly proportional to the growth rate. It is noted that the ratio is rather constant for high growth rates but that the assimilation is faster as the growth rate of the population before the addition of carbon was low. Care should be taken to recognize that in the steady state, before the perturbation, the ratio R corresponds to the yield of conversion of carbon into biomass. This quantity is globally independent of the growth rate (dashed line in Figure 3.9).

Further illustrations of this phenomenon can be found in the literature. Leegwater *et al.* [LEE 82], Lendenmann and Egli [LEN 98], Natarajan and Srienc [NAT 99, NAT 00] and Lara *et al.* [LAR 09] observed the same phenomenon with different measurement methods. The results of Leegwater *et al.* are shown in Figure 3.10. The experiments carried out consist of a rapid addition of substrate to a chemostat culture for different dilution rates and different strains. At the steady state, the assimilation rate is proportional to the growth rate (black line). Following the addition of sugar, the measured rate (green dot) exceeds the measured value in the steady state. The authors also measured *in vitro* the rate of assimilation due solely to the PTS transport system.

It should be noted that the instantaneous value of the assimilation rate exceeds that of the PTS system (orange). Hence, we can conclude that another transport system is activated under a limiting condition. We observe, however, that the behavior varies from one strain to another. In *E. coli*, the contribution of transport systems other than PTS is higher at lower growth rates. For *Klebsiella aerogenes*, on the contrary, it appears that carriers other than PTS are active at high growth rates. The work of Sweere *et al.* on *Saccharomyces cerevisiae* yeast also shows de-coupling between growth rate and assimilation rate in the transient regime [SWE 88, SWE 89].

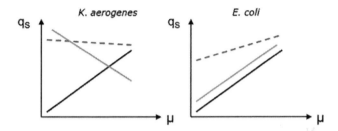

Figure 3.10. *Measured value of assimilation rate as a function of growth rate, in the steady state (continuous line), after addition of substrate (- - -) and due to the PTS system alone in vitro (orange). According to Leegwater et al. [LEE 82]. For a color version of this figure, see www.iste.co.uk/morchain/bioreactor.zip*

These datasets thus confirm the existence of several transport systems, mobilized in a variable way according to the residual concentration in the medium and the type of strain. The maximum assimilation rate measured in a batch reactor is not a true biological constant. The maximum assimilation rate also depends on the nature, the number of transport systems and their respective maximal activity. All these quantities are dependent on the type of culture (especially the feed concentration for a chemostat) and the cells' history. Thus, Lara *et al.* measured the instantaneous assimilation rate achieved by *E. coli* bacteria previously cultured as a chemostat. The results show an instantaneous assimilation capacity of 10–30 times greater than the assimilation rate observed in the steady state. This over-assimilation manifests itself almost instantaneously (less than a second) but does not extend beyond 20 s.

Mode	$q_S(mmol.g_X^{-1}.h^{-1})$	$q_{O_2}(mmol.g_X^{-1}.h^{-1})$
Chemostat ($\mu = 0.1\ h^{-1}$)	1.35	3.78
Aerobic Pulse	15 ± 1	16
Anaerobic Pulse	31 ± 3	-

Table 3.1. *Specific assimilation rates of glucose and oxygen during exposure of E. coli to a pulse of glucose with and without supplemental oxygen supply. According to Lara et al. [LAR 09]*

3.3.3. *Dynamic response of the metabolism*

From a technical viewpoint, this type of study requires considerable expertise using either fluorescent markers [DEL 10, DEL 11, SUN 12], metabolites or radioactive isotopes [MAR 12, NÖH 07, WIE 02].

The experimental information in this field is rich but complex. It relates to both assimilation rates and the production rates of intra- and extra-cellular metabolites. Indeed, the responses of the cell can be manifested in many places from the gene to the production of metabolites and thus on very different time scales. Pre-culture conditions, population state at the time of the imposed perturbation and the nature of this perturbation play a significant role in interpreting the results [LOO 05]. It is not surprising to find conflicting conclusions by different teams that have carried out identical studies. In this context, work on characteristic times is very useful for organizing and classifying experimental information [ESE 83, KRE 04].

Figure 3.10 illustrates the main devices used in the laboratory to perform such studies.

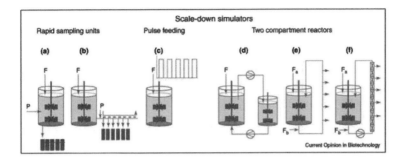

Figure 3.11. *Laboratory reactors for the "simulation" of degraded macromixing conditions in industrial bioreactors. Taken from Neubauer and Junne [NEU 10]*

This complexity in the realization and analysis of the results is illustrated in an experiment designed to mimic concentration variations seen by cells in a large bioreactor: prolonged exposure to low concentrations interspersed with sudden and repeated exposures to much higher concentrations. For this purpose, the experimental device (f) of Figure 3.11 is used.

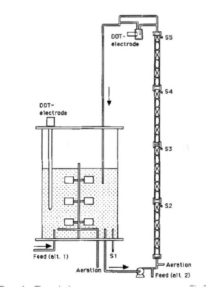

Figure 3.12. *Diagram of the "scale-down" reactor allowing the periodic exposure of cells to excess substrate when passing through the system's plug flow zone [NEU 95]*

The cells travel through the loop consisting of a stirred reactor (STR) and a plug flow reactor (PFR). The residence time in the plug flow reactor is 2 min. The mean residence time in the stirred reactor was set at 27 min using a pump. After a batch culture phase, the system is switched to fed-batch and a progressive decrease in the growth rate is imposed throughout the experiment (8 h). The sugar supply made at the bottom of the plug flow reactor periodically exposes the cells to a high concentration[4]. The sugar concentrations are measured along the plug flow reactor, which makes it possible to evaluate the assimilation rates of the cells over a time scale of about 100 s. The excretion of metabolites such as acetate and formate is also measured as shown in Figure 3.13.

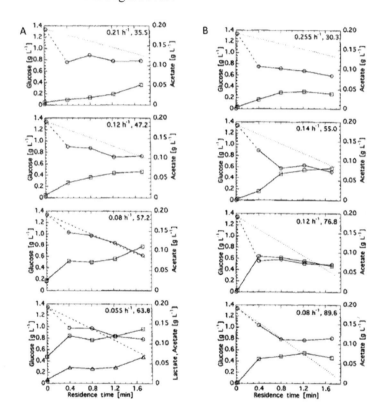

Figure 3.13. *Concentrations measured in the plug flow reactor in the absence (A) or presence (B) of oxygen-enriched air. The values at the top right corner of each graph indicate the growth rate and the cell concentration at a given time [NEU 95]*

---

4 By contrast, feeding in the stirred reactor limits these over-concentrations because the rate of mixing is high and the volume is large.

By analyzing the results of the work of Neubauer *et al.* already mentioned in section 3.2.2.5 devoted to transporters, we will be able to observe the consequences of the disequilibrium between the assimilation rate and growth rate. By entering the plug flow reactor, the cells are always exposed to the same concentration. It is observed in these graphs that the assimilation rate of the glucose at the inlet of the plug flow reactor exceeds the maximum value identified in the batch reactor (the expected concentration decrease based on this value is indicated by the dotted line). The difference between these two rates is reduced as the experiment progresses.

Under transient conditions, the instantaneous assimilation rate is thus not directly related to the growth rate. It changes rapidly in about 30 s, giving a characteristic time in the order of 5–6 s. It is also observed that this dynamic adaptation involves a complex regulation, in which the available oxygen flow plays an important role.

In any case, this over-assimilation of glucose results in the production of acetate. This metabolite is produced when the flux of assimilated sugar is in excess with respect to the usage capacity of the cell. We also observe a blockage of the assimilation in a few tens of seconds, which can be attributed to either inhibition by the acetate or a retro-action mechanism following over-assimilation. Two quite different responses are observed depending on the presence or absence of oxygen in the medium.

– If the gas is not enriched with oxygen, acetate concentrations increase over time and the final quantity produced increases with the amount of cells present (we note that the growth rate decreases concomitantly).

– If the gas is enriched with oxygen, the total production is practically independent of the growth rate. The concentrations stagnate or even decrease as soon as the sugar's assimilation rate becomes nil.

The analysis is all the more difficult when several dynamics are at work: the population's specific growth rate and the glucose residual concentration in the mixed reactor decrease at the same rate, and the assimilation system plays a role on shorter time scales just like the metabolic one. Finally, there is a distribution frequency of passage through the piston induced by the distribution of the residence times in the stirred reactor.

Figure 3.14 shows a summary of the obtained results. The main experimental fact is that after a period of nutritional limitation (famine) the overall assimilation rate during a sugar addition is very much higher than the maximum assimilation rate measured during the growth phase in batch mode, that is to say, not limited by glucose. Masses of 2–15 g of sugar per gram of cell per hour ($g_S.g_X^{-1}.h^{-1}$) are obtained against a maximum of 2 $g_S.g_X^{-1}.h^{-1}$ in a batch reactor.

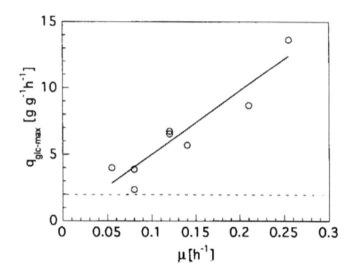

Figure 3.14. *Initial value of the specific assimilation rate following a glucose pulse. Fed-batch E. coli culture grown in sugar limitation. The horizontal line indicates the maximum value measured in continuously non-limiting conditions (Batch culture) [NEU 95]*

CONCLUSION.– It can thus be concluded that the cells modulate the activity of the transporters dynamically over relatively short time scales of the order of the second, that is to say, similar to the macroscopic mixing time (even in small laboratory reactors). This modulation is coupled with metabolism and has energetic aspects as shown by the role played by oxygen.

Again, as Kremling *et al.* [KRE 04] pointed out, the measured assimilation rates depend in part on the population's state, that is, the culture's conditions and how the perturbation is applied. Thus, an important task that needs to be carried out is the collection and synthesis of the different values measured in relation to the particular conditions of each study.

3.4. Equilibrium models, dynamic models

Biological kinetic description models are usually distinguished according to two criteria: use of intracellular quantities or not (structured or unstructured model) and distinction between different individuals or not (population or average individual model). We propose here to distinguish according to an objective and useful criterion for considering the coupling between the transport equation and the equations describing the biological system's dynamics: the characteristic time(s) of the biological phenomena considered in the model. This methodology has traditionally been used in process engineering and fluid mechanics, where it has proved itself, and all indicates that it will be the case here.

The first benefit of this approach lies in an elementary principle to which we must conform: everything that is not simulated must be modeled. Typically, it will be for any modeling to define a characteristic time scale. All the physical and biological phenomena whose characteristic times are larger than the reference scale will have to be simulated. In order to account for phenomena whose characteristic time is smaller than the reference scale, modeling will be proposed. In general, we speak of a model for closing the equations. This closure can be based on a strongly time-resolved simulation, on a compilation of experimental data integrated in correlation form or on a mathematical proposition (a closure model) whose relevance will be judged by comparison with experimental data. A model will therefore be both dynamic for the scales solved by simulation and at equilibrium with respect to the modeled scales.

Below the chosen reference scale, the dynamics of rapid phenomena are therefore not explicitly described but integrated into the equations to take account of the observable effects at the reference scale. By way of illustration, consider Pirt's law. This law describes "on the growth's scale" the fact that intracellular content and kinetics change with respect to the growth rate, which affects the conversion yield of the substrate into biomass. When using this law in a simulation modeling approach, we do not attempt to describe how, why or how fast the functioning of microorganisms changes, but we do integrate this reality in the form of a variable conversion yield relative to the growth rate.

3.4.1. *Unstructured kinetic model: equilibrium model or zero-equation model*

The unstructured model describes the biomass without reference to its composition. It is similar to the chemical kinetics model in homogeneous phase. These models allow a description of most of the fundamental phenomena of bacterial growth. They are based on the assumption of a balanced growth during which the composition of the biomass remains constant. It is a model where all small-scale biological phenomena are filtered (averaged) and provide information about the growth scale.

An unstructured kinetic model expresses the rate of the transformations in the form of specific velocities (grams per gram of cell and per hour) according to a set of equations of type [3.22]:

$$r_X = \mu X \qquad \text{Biomass growth}$$

$$r_S = -\frac{\mu}{Y_{SX}} X \qquad \text{Substrate consumption}$$

$$r_{O_2} = -\frac{\mu}{Y_{OX}} X \quad \text{Respiration} \qquad\qquad [3.22]$$

$$r_p = \frac{\mu}{Y_{PX}} X \qquad \text{Formation of product P}$$

Kinetic models are considered simple and easy to implement. The set of rates are determined from the specific growth rate (μ) and the conversion yield (grams of product consumed or formed per gram of cell). The specific growth rate is generally given by Monod's equation [MON 49]. This reflects the effect of limiting substrate concentration on the growth rate of the microorganism:

$$\mu = \mu_{max} \frac{S}{K_S + S} \qquad\qquad [3.23]$$

where

– μ_{max}: maximum specific growth rate without limiting factor (h^{-1});

– K_S: saturation constant ($g.l^{-1}$).

Monod's model allows modeling only phases of growth and slowdown. It is based on the assumption that the growth rate is limited[5] by the enzymatic kinetics of the substrate's metabolism. The formulation of Monod's law has been extended to the case of several substrates such that the microorganisms' specific growth rate is related to the concentrations in the liquid phase according to equation [3.24]. The overall rate is obtained by taking the product of the hyperbolic functions, which reflects the fact that all the involved substrates are necessary to achieve growth:

$$\mu = f(\mathbf{C_L}) = \prod_{1 < i < n} \frac{C_i}{K_i + C_i} \qquad [3.24]$$

In order to describe the fact that a microorganism has several growth modes based on different chemical compounds and/or metabolisms, a summation is used. It is important to note that the rate in each mode can be described by a law of the form [3.24]:

$$\mu = \alpha(\mathbf{C_L})\mu_1 + (1 - \alpha(\mathbf{C_L}))\mu_2 \qquad [3.25]$$

The function $\alpha(C_L)$ is used to control switching between growth modes. Water treatment models, such as the Activated Sludge Model, or waste treatment models, such as the Anaerobic Digestion Model, are based on this formalism.

CRITICAL ANALYSIS.– The main advantage of unstructured models lies in their simplicity and ease of implementation. By construction, the characteristic time scale of these models is of the order $\tau_{NS} = 1/\mu_{max}$. These models are integrated at the growth scale.

They rely on an algebraic relationship between liquid-phase concentrations and growth rate. The notion of dynamics of the biological system is absent and we can therefore speak of a model of equilibrium or of a model with zero equation, as no equation accounts for the dynamic evolution of the properties of the biological system.

Although this model is particularly suited to the monitoring of bioreactors under steady-state and pseudo-steady-state conditions, it proves to be inoperative when studying a reactor under dynamic conditions. Indeed,

---

5 See also the citation of C. J. Perret at the end of this section.

numerous studies show differences and contradictions between unstructured kinetic models and experimental facts in the case of microorganisms subject to sudden variations in their environment [MOR 09, SIL 08]. Moreover, because these models make no distinction according to the internal composition of the cells, they describe average behavior at the level of the population. These models, as explained by Nielsen and Villadsen [NIE 92], are inoperative when the internal structure of the cells plays an important role in the studied response or when the cells are confronted to sudden variations in their environment. It is illusory and unnatural to attempt to describe rapid biological dynamics with this type of models. Their field of application is thus limited to the study of homogeneous bioreactors in the steady state or pseudo-steady state.

NOTE.– There are unstructured kinetic models with variable yields. In this case, yields are expressed as functions of growth rate and indirectly to the substrate concentration.

The most common formulation in this case is by using the so-called "Pirt" law:

$$\frac{1}{Y_{SX}} = \frac{1}{Y_{SX,max}} + \frac{m}{\mu} \qquad\qquad [3.26]$$

Without going into detail, variable coefficients reflect an underlying modification of the metabolism. They reflect these changes at the growth scale.

3.4.2. *Structured kinetic model*

Structured models incorporate information related to the biological phase. This information may relate to the composition, rate of intracellular processes and physiology. The introduction of variables specific to the biological phase makes it possible to distinguish between individuals[6]. For the time being, we will stick to kinetic models on the basis of the notion of the average individual.

The conservation equation of an intracellular component c_i in an open reactor is given by equation [3.27]:

---

6 We enter into the field of populations, which will be treated separately in Chapter 4.

$$\frac{d\left(c_i XV\right)}{dt} = R_i(\mathbf{c})XV - Qc_i X \qquad [3.27]$$

where $R_i(\mathbf{c})$ represents the net specific rate for the production of c_i related to all the intracellular reactions.

With the notations conventionally used in chemical reaction engineering, this rate will be expressed using the expression [3.28]:

$$R_i(\mathbf{c}) = \sum_j v_{i,j} r_j(\mathbf{c}) \qquad [3.28]$$

where $v_{i,j}$ is the stoichiometric coefficient for the compound i engaged in the reaction j. It is recalled that this coefficient is positive if the compound c_i is produced by reaction i and negative if the compound c_i is consumed by reaction j.

r_j denotes the reaction rate j. In the case of intracellular biological reactions, this is most often a Michaelian kinetic expression:

$$r_j\left(c_i\right) = r_{j,\max} \; e_j \frac{c_i}{k_i + c_i} \qquad [3.29]$$

Let us now see how to transform the conservation equation under the hypothesis of a constant-volume reactor:

$$VX \frac{dc_i}{dt} + Vc_i \frac{dX}{dt} = R_i(\mathbf{c})XV - Qc_i X \qquad [3.30]$$

Then, dividing on both sides by the product VX:

$$\frac{dc_i}{dt} + c_i \frac{1}{X} \frac{dX}{dt} = R_i(\mathbf{c}) - \frac{Q}{V} c_i \qquad [3.31]$$

Finally, we get equation [3.32]:

$$\frac{dc_i}{dt} = R_i(\mathbf{c}) - \mu.c_i - \frac{Q}{V} c_i \qquad [3.32]$$

where μ is the specific growth rate.

Thus, the second term in the right-hand side is seen as a dilution term for the mass of c contained in the cells because of their multiplication. In the absence of the production of c, the initial mass decreases exponentially as the cells divide. It is analogous to the last term, which is hydrodynamic in nature. This model therefore uses three time scales, two of which are biological: one linked to the intracellular reaction rate and the other linked to growth. The third time scale is given by the mean residence time.

Let us return now to the passage from equation [3.31] to equation [3.32]. We had to place $\frac{1}{X}\frac{dX}{dt} = \mu$. This proposition is based on the assumption of a balanced growth for which the multiplication in number (exact definition of μ) corresponds to a mass multiplication, which basically amounts to positing that the mean density of the cells remains constant at the growth scale. Let us take again equation [3.31], assume that the reactor is now closed and integrate over a long time period in relation to the growth time scale:

$$\frac{1}{T}\int_t^{t+T}\left(\frac{dc_i}{dt} + c_i\frac{1}{X}\frac{dX}{dt}\right)dt = \frac{1}{T}\int_t^{t+T} R_i(\mathbf{c})dt \qquad [3.33]$$

Over a long period of time and for a compound i forming part of the cell's structure, it can be assumed that the cell's average composition does not vary:

$$\frac{1}{T}\int_t^{t+T}\frac{dc_i}{dt}dt = 0 \qquad [3.34]$$

We can therefore introduce the mean value $\overline{c_i}$ into equation [3.33]:

$$\frac{\overline{c_i}}{T}\int_t^{t+T}\frac{\dot{X}}{X}dt = \overline{R_i(\mathbf{c})} \qquad [3.35]$$

Hence, the formulation of the evolution of the cell mass over a long period of time is given by equation [3.36]:

$$X(t+T) = X(t)\; e^{\frac{\overline{R_i(\mathbf{c})}}{\overline{c_i}}T} \qquad [3.36]$$

It is therefore possible to relate the average specific growth rate over the period T to $\overline{R_i(C)}/\overline{c_i}$, that is to say, the mean specific synthesis rate of the main constituents of the cell. The passage from equation [3.31] to equation [3.32] is justified only for a long time, considering that there is no accumulation (equation [3.34]). Apart from this hypothesis, and particularly for compounds involved in storage phenomena, it is not advisable to simplify equation [3.23].

The addition of new variables subject to conservation laws (composition) introduces new time constants into the model, which then acquires a real dynamic character. However, this improvement is achieved at the expense of a significant increase in the model's complexity. Moreover, it is difficult to monitor a large number of metabolites for several reasons:

– the number of intracellular metabolites is very large;

– the number of possible reactions is also very large;

– the kinetics of these reactions are only partially known;

– the number of parameters to be adjusted is considerable.

Some intermediates are present in trace quantities, not detectable experimentally, which complicates the model's validation.

Thus, although it is interesting to attempt to establish a model describing the set of intracellular compounds, the task is in practice unrealizable. Yet this model would have the advantage of explicitly showing the set of characteristic times associated with all biological processes.

In the absence of the possibility of carrying out an approach that could be described as a direct simulation (describing all the compounds and therefore all the scales through complete mass and energy balances), reduced models were formulated. The concept of a reduced model has two aspects: that of a model constructed immediately from a restricted number of variables and that of a model simplified from a more complex model (model reduction approach).

The idea used in this type of model is to consider the existence of compartments inside the cells and to distribute the biomass between these compartments. Each compartment is the site of a specific cellular activity. This will involve, for example, the assimilation of a substrate, the production of a product or the synthesis of another compartment. Thus, some compartments contribute to the growth of biomass while others control the production of enzymes necessary for growth. This type of model requires good knowledge in biochemistry and microbiology in order to identify the main metabolic pathways as well as intracellular chemical species involved in metabolic reactions (enzymes, ATP, ADP, etc.) [NIE 92].

CRITICAL ANALYSIS.– Structured models describe the evolution of the microorganisms' composition or properties. They provide a considerable improvement in the description of the problem by making a distinction between the liquid-phase compounds and the intracellular compounds. Moreover, they consider the existence of several characteristic time scales.

On the contrary, the number of solved variables increases and the time scale characteristic of the model corresponds to the time scale of the fastest biological phenomenon considered. The association of these structured models with a microorganism's transport model poses problems because when two contained currents of the cells are mixed, the liquid is homogenized but not the cells' contents. In the new environment, each individual sub-group follows a specific trajectory related to its state (see Conclusion of Chapter 2).

3.4.3. *Metabolic model*

Unlike kinetic models, metabolic models do not postulate the existence of a constant yield. They calculate the specific consumption rates (substrate and oxygen) and production rates (metabolites and biomass) from the fluxes of the substrates entering the cell and the mapping of the paths that can be taken by the substrates in the cell. We call the latter metabolic networks or metabolic pathways. The yield of each reaction is known but the distribution of material flows in the different channels results in overall yields varying as a function of the cell's feed conditions. Thus, the growth yield, grams of biomass formed per gram of substrate, is a result of the model. Therefore, the metabolic models make it possible to represent the different modes of the cells' functioning. However, as for kinetic models, metabolic models invoke a hypothesis of non-accumulation within microorganisms. They are based on writing material balance and conservation of energy and ignore the phenomenon of accumulation within biomass. Results are highly dependent on:

– the choice of reactions and their organization (metabolic network);

– the objective function used to solve the underlying optimization problem;

– the quality of the measurements of the fluxes exchanged with the environment.

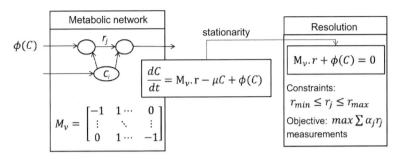

Figure 3.15. *Principle of a metabolic model based on stoichiometry. At the growth scale (over a long period of time), the hypothesis of non-accumulation (or stationarity of the composition) leads to a linear system of equations dealing with rates. Generally under-determined, this system is solved via an under-constrained optimization procedure. From Llaneras and Picó [LLA 08]*

Among the metabolic models, we can distinguish:

– Metabolic flux models that aim to identify fluxes in each branch of the network. In general, several solutions (or models) are acceptable as might be expected for a strongly coupled reaction system [CHU 96]. It is tempting to think of experimental observation as the superimposition of several models between which cells move dynamically [CAR 04]. These models are not strictly dynamic insofar as the distribution of the fluxes is re-calculated for each new environmental condition without taking into account the previous state. The objective of these models is rather the identification of the possible paths taken by the material. This identification is largely based on measurements of consumed and excreted fluxes.

– The metabolic/cybernetic models [JON 99], which combine a structured approach for the description of key enzyme intracellular concentrations and the so-called cybernetic variables controlling the rate of synthesis of those key enzymes. This approach makes it possible to simplify the representation of all the mechanisms of the cell's regulation. To solve the problem, we add objective functions to the cybernetic variables (maximization of growth, minimization of energy expenditure). With these models, the concentrations of the reaction intermediates are calculated.

– Another model consists of describing the metabolic reaction rates and defining a probability distribution for the value of the coefficients that appear in these rate laws [HEI 05]. This type of model forms the basis of the population models called "cell ensemble models".

It must be understood that the basis of the problem in modeling comes from the fact that the metabolic network is not totally fixed: the cell is able to change its metabolism either through the modulation of the enzymes' activity controlling a reaction or by inducing new metabolic pathways. This network plasticity obviously poses major problems in terms of modeling and yet this is where the real need for industrial bioprocesses lies [VAR 99].

The key distinction between conventional chemical reaction systems and metabolic networks, which is often missing in kinetic metabolic network models, is the influence of regulation and control. In conventional chemical reaction systems, knowledge of the kinetics completes the treatment of the system. In biological systems, however, all levels of metabolic function (i.e. transcription, translation, and catalytic activity) are tightly integrated and coordinated with the global environment of the organism, hence yielding adaptability in the face of changing conditions. Thus, the 'conventional' mathematical treatment of a metabolic network, encompassing only the kinetics and stoichiometry, is often hard pressed to correctly predict system adaptation because it lacks a description of the forces driving the adaptation. This capability, however, is exactly what is required for the reengineering of physiology. [VAR 99].

3.5. Confrontation of models with experimental data

Some examples from the literature are discussed in order to clarify these notions of dynamics, adaptation and equilibrium. It will not be a question of fitting the parameter of a model but of analyzing the functioning of the main types of models in selected configurations. To this end, we reiterate the cases put forward in the study of experimental data.

3.5.1. *Response of a chemostat to an increase in the dilution rate*

3.5.1.1. *Zero-equation model*

We formulate a very simple approach involving an unstructured biological model. The bioreactor can be described using equation set [3.37]. In order to simplify, we assimilate the medium's volume to the liquid's volume, which is consistent with the low volume fraction in cells ($X = 5$ g.l^{-1} equals $\varepsilon_S = 0.5\%$):

$$\frac{dX}{dt} = \left(-\frac{Q_L}{V} + \mu\right)X$$

$$\frac{dS}{dt} = \frac{Q_L}{V}(S_e - S) - \frac{1}{Y_{SX}}\mu X \qquad [3.37]$$

$$\mu = f(S) = \mu_{max}\frac{S}{K_S + S}$$

At the steady state, the well-known identity between the mean growth rate[7] of the cells μ and the dilution rate Q_L/V is obtained.

If the above system of equations is used to calculate the dynamic response to a sudden change in the dilution ratio, we obtain the results presented in Figure 3.16.

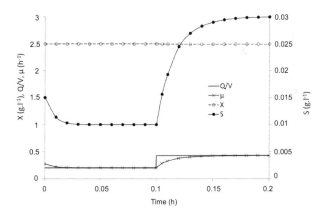

Figure 3.16. *Simulation of the response to a step-up on the dilution rate (Q/V = 0.2– 0.4825 h−1) with a zero-equation model. μmax = 1 h−1, KS = 40 mg.l−1, Se = 5 g.l−1*

---

7 This is an average taken overall individuals.

The results obtained by solving the set of equations [3.9] indicate that the cell concentration does not change following a flow rate change Q_L and that the substrate concentration adjusts very quickly to its new value (as does the growth rate algebraically related to it: $\mu = f(S)$). Moreover, the concentration variation is small, the value increases from 10 to 30 mg.l^{-1} and remains in the order of magnitude of the constant K_S. These results are comparable to those presented previously in Figure 3.4 and correspond to the expected response in the case of a small perturbation.

These dynamics can be analyzed in terms of linear systems. To do this, we start from the expression of the evolution of the substrate concentration:

$$\frac{dS}{dt} = \frac{Q_L}{V}\left(S_e - S\right) - \frac{1}{Y_{SX}}\mu_{max}\frac{S}{K_S + S}X \tag{3.38}$$

Let us linearize this expression around the steady state that precedes the change in the dilution rate. It is a question of expressing the variation of $S(t)$ with respect to its value before perturbation $\overline{S}$. This variation is noted $\tilde{S}(t)$. We use a Taylor expansion of a function of multiple variables:

$$\frac{dS}{dt} = f(Q_L, S, X) \Rightarrow \frac{d\tilde{S}}{dt} \approx \sum\left(\frac{\partial f}{\partial x_i}\right)_{ST}\tilde{x}_i \tag{3.39}$$

In this formula, the partial derivatives are evaluated from the values of variables in the steady state (hence the ST index). The latter are by convention surmounted by a bar. By neglecting the variation in the biomass concentration, we get the following calculation:

$$\frac{d\tilde{S}}{dt} = \frac{\tilde{Q}_L}{V}\left(S_e - \overline{S}\right) - \left(\underbrace{\frac{1}{Y_{SX}}\frac{\mu_{max}K_S}{\left(K_S + \overline{S}\right)^2}\overline{X} + \frac{\overline{Q}_L}{V}}\right)\tilde{S} \tag{3.40}$$

We are dealing with a linear system of order 1. The underlined term is of dimension inverse to time. This time corresponds to the characteristic time of the concentration's response to a variation of the flow rate Q_L. A numerical application leads here to $\tau = 0.0125$ h. This response time is very short, less than a minute. Given that equation [3.40] corresponds to a linear system of order 1, we can deduce that the concentration S will not vary after a period of time equal to approximately five times the characteristic time τ, that is, 0.0625 h, approximately 4 min. This estimate is quite consistent with what is predicted by the simulation.

By adding a Pirt's law-like equation $Y_{SX} = f(\mu)$, we can improve the prediction of the concentration of microorganisms and, in particular, account for the fact that the final biomass concentration is higher. The results obtained with the set of equations [3.37] and variable yields are presented in Figure 3.17.

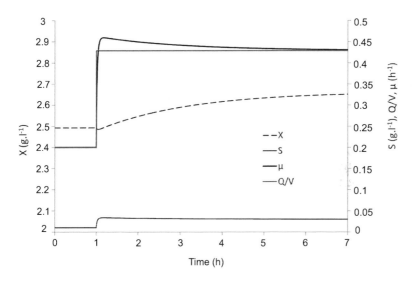

Figure 3.17. *Simulation of the response at a moderated scale over the dilution rate. Zero-equation model with variable conversion yield. Parameters:* $\mu_{max} = 1\ h^{-1}$, $K_S = 40$ $mg.l^{-1}$, $S_e = 5\ g.l^{-1}$. $Y_{SX,max} = 1.75\ et\ m = 0.05$

With the addition of a variable yield, a behavior similar to that depicted in Figure 3.4 is observed, in particular with a higher terminal microorganism concentration in connection with a higher growth rate. A very slight decrease in the concentration of microorganisms is observed just after the increase in flow rate. With the numerical integration being carried out with a time step of 5×10^{-5} h or (0.2 s), we can rule out the hypothesis of a numerical artifact. This phenomenon of very small amplitude and short duration is practically and experimentally undetectable.

The modification of the dilution ratio here is moderate: $\Delta D/\mu_{max} \approx 0.28$. If we impose a much higher variation $\Delta D/\mu_{max} \approx 0.7$, we obtain the results presented in Figure 3.18, which illustrate the behaviors with a constant and variable yield with respect to the specific growth rate.

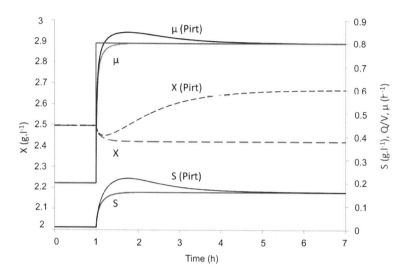

Figure 3.18. *Response to a significant increase in the dilution rate. Results of zero-equation models with constant and variable yields (Pirt). The parameters are identical to the previous cases*

A significant increase in the dilution rate results in an increase in the residual substrate concentration and a decrease in the concentration of microorganisms with a constant yield model. In addition, the growth rate adjusts to the new dilution rate in just 30 min. If a variable growth rate is used, stabilization is slower. The general trend seems to coincide with experimental observations but the amplitude is significantly underestimated: the substrate concentration only slightly exceeds the final value and the biomass decay remains small and circumscribed to 30 min following the perturbation. Compared with experimental data, examination of Figure 3.5 shows that the zero-equation model, or instantaneous equilibrium model, does not describe the response to large variations in the dilution rate.

3.5.1.2. *One-equation model*

Many proposals have been made to establish a biological model reflecting the growth rate's adjustment dynamics. In a first step, it was a question of improving an unstructured kinetic model by establishing a relationship between the variations of the substrate concentration and the variations of the growth rate [YOU 70]. The following cited article is particularly relevant [YOU 70]:

A number of experimental studies on the dynamic behavior of the chemostat have shown that the specific growth rate does not instantaneously adjust to changes in the concentration of limiting substrate in the chemostat following disturbances in the steady state input limiting substrate concentration or in the steady state dilution rate. Instead of an instantaneous response, as would be predicted by the Monod equation, experimental studies have shown that the specific growth rate experiences a dynamic lag in responding to the changes in the concentration of limiting substrate in the culture vessel. The observed dynamic lag has been recognized by researchers in such terms as an inertial phenomenon and as a hysteresis effect, but as yet a system engineering approach has not been applied to the observed data.

The approach used to establish a dynamic model valid for a chemostat is system based. Instead of considering an algebraic relationship between growth rate and substrate concentration, it postulates a general relationship and further tries to identify its exact nature. It considers the fluctuations with respect to a steady state and goes through a step of linearization of the basic equations around the steady state preceding the imposed external perturbation. Let us start by re-writing the equations for a chemostat revealing the dependencies between variables:

$$\frac{dX(t)}{dt} = \left(-D(t) + \mu(S,t)\right)X(t)$$

$$\frac{dS(t)}{dt} = D(t)\left(S_e(t) - S(t)\right) - \frac{1}{Y_{SX}}\mu(S,t)X(t)$$

[3.41]

These equations can then be linearized around a steady state. To do this, we can choose to:

– write each variable as the sum of an average component and a fluctuation: $y(t) = \bar{y} + \tilde{y}(t)$ then inject this into the equations and neglect the terms involving the product of two fluctuations;

– perform a Taylor expansion of the balance equations.

Let us apply the latter method to the system's equations [3.41], noting that the relationship between μ and S is not known. In order to establish a general formulation, we consider here that quantities S, S_e, X and μ are time dependent:

$$\frac{d\tilde{X}(t)}{dt} = \left(\bar{\mu} - \bar{D}\right)\tilde{X}(t) + \bar{X}\,\tilde{\mu} - \bar{X}\,\tilde{D}$$

$$\frac{d\tilde{S}(t)}{dt} = -\left(\bar{D} + \frac{\bar{X}}{Y}\left(\frac{\partial\mu}{\partial S}\right)_{RP}\right)\tilde{S}(t) + \left(\bar{S}_e - \bar{S}\right)\tilde{D}(t) - \frac{1}{Y_{SX}}\bar{\mu}\tilde{X}(t) + \bar{D}\tilde{S}_e(t)$$

[3.42]

The following constants are now defined:

$$\tau_H = \frac{1}{\bar{D} + \dfrac{\bar{X}}{Y}\left(\dfrac{\partial\mu}{\partial S}\right)_{ST}}, \quad \tau_D = \frac{1}{\bar{D} - \bar{\mu}}$$

[3.43]

$$G_X = \frac{\bar{\mu}}{Y_{SX}}\tau_H, \quad G_{Se} = \bar{D}\tau_H, \quad G_\mu = \bar{X}\tau_D, \quad G_D = \left(\bar{S}_e - \bar{S}\right)\tau_H$$

which leads us to the expression of system [3.44] in canonical form:

$$\tau_D\frac{d\tilde{X}(t)}{dt} + \tilde{X}(t) = G_\mu\,\tilde{\mu}(t) - G_\mu\,\tilde{D}(t)$$

$$\tau_H\frac{d\tilde{S}(t)}{dt} + \tilde{S}(t) = -G_X\tilde{X}(t) + G_{Se}\tilde{S}_e(t) + G_D\tilde{D}(t)$$

[3.44]

We can then apply a Laplace transformation to the system of equations [3.44] to obtain equations [3.45]:

$$\tilde{X}(p)\left(1 + \tau_D p\right) = G_\mu\,\tilde{\mu}(p) - G_\mu\,\tilde{D}(p) = G_\mu\,F(p)\tilde{S}(p) - G_\mu\,\tilde{D}(p)$$

$$\tilde{S}(p)\left(1 + \tau_H p\right) = -G_X\tilde{X}(p) + G_{Se}\tilde{S}_e(p) + G_D\tilde{D}(p)$$

[3.45]

The corresponding block diagram is shown in Figure 3.19.

In order to establish a possible form for the transfer function $F(p)$, Young and colleagues relied on a mass balance for an intracellular compound s, considering assimilation, excretion and use of this compound by intracellular reactions. By postulating that the transport mechanisms obey Michaelian kinetics, they derived equation [3.46]:

$$\frac{ds(t)}{dt} = r_{S,up,max} \frac{S(t)}{k_{up} + S(t)} - r_{S,ex,max} \frac{s(t)}{k_{ex} + s(t)} - r_{max} \frac{s(t)}{k_s + s(t)} \qquad [3.46]$$

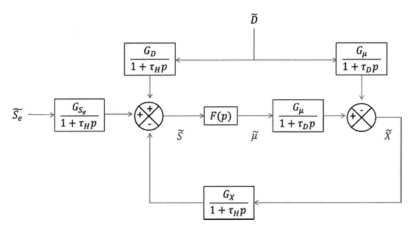

Figure 3.19. *Block diagram of a chemostat. The transfer function F(p) is to be identified from the experimental data*

The last term is referred to as the dynamic growth rate μ(t). After linearization and adding the fact that the induction of enzymes (or the activation of an enzymatic pool) allowing an increase of the growth rate requires a certain time, the authors proposed equation [3.47], in which τ_{ind} represents the time delay related to the induction:

$$F(p) = \frac{K \ e^{-\tau_{ind}p}}{1 + \tau_s p} \qquad [3.47]$$

where τ_s corresponds to the characteristic time of the evolution of the intracellular concentration under the effect of reactions and excretion:

$$\tau_s = \frac{1}{r_{S,ex,max}/k_{ex} + r_{max}/k_s} \qquad [3.48]$$

Such a model is capable of qualitatively describing a wide range of responses to changes in input concentration or dilution rate. Figure 3.20 shows that the authors obtained satisfactory results for the response of a chemostat to a sudden increase in substrate concentration (assuming here that $\tau_{ind} = 0$).

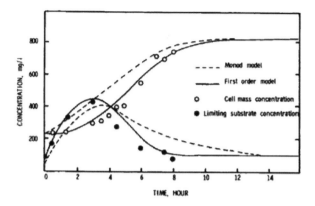

Figure 3.20. *Simulation of a chemostat's response to a sudden increase in feed concentration ($\tau_{ind} = 0$ in equation [3.47]). Data from Storer and Gaudy [STO 69]*

On the contrary, the agreement is somewhat less convincing for the response to a shift of the dilution rate, as shown in Figure 3.21 despite a different adjustment of the transfer function ($\tau_s = 0$ and $\tau_{ind} \neq 0$).

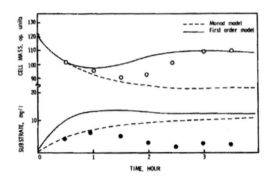

Figure 3.21. *Simulation of a chemostat's response to a sudden increase in the dilution ratio. $\tau_{ind} = 0$ in equation [3.47]). Data from Mateles et al. [MAT 65]*

In Young *et al.*'s model, the internal variable, *s*, was finally eliminated, but other proposals emerged in the form of structured kinetic models. Typically, a differential equation for an intracellular concentration analogous to the external carbon source is introduced into the model and the growth rate is related to this intracellular variable [ABU 89, PAT 00]:

$$\frac{dX(t)}{dt} = \left(-D(t) + \mu(w)\right) X(t)$$

$$\frac{dS(t)}{dt} = D(t)\left(S_e(t) - S(t)\right) - \frac{1}{Y_{SX}} \mu(w) X(t) \qquad [3.49]$$

$$\frac{dw(t)}{dt} = \frac{1}{\tau_w}\left(S(t) - w(t)\right)$$

$$\mu = f(w)$$

These models introduce the idea that the specific growth rate constitutes a property of the microorganisms defined in relation to the composition, w, of the latter. A Monod-type law is used to relate the instantaneous growth rate to the internal variable w. For further details, see Silveston *et al.* [SIL 08].

Rather than referring to an internal variable, w, with a fuzzy definition and for which it is not easy to establish an evolution equation, we proposed to construct a model describing the evolution of the population's average growth rate [MOR 00, MOR 09]:

$$\frac{dX}{\partial t} = \left(-D + \mu\right) X$$

$$\frac{dS}{dt} = D\left(S_e - S\right) - \frac{1}{Y_{SX}} \mu X$$

$$\frac{d\mu}{dt} = \frac{1}{\tau_\mu}\left(\mu^* - \mu\right) \ \ with \ \ \mu^* = f(S) \qquad [3.50]$$

or

$$\frac{d\mu}{dt} = \left(\frac{1}{\tau_\mu} + \mu\right)\left(\mu^* - \mu\right) \ \ with \ \ \mu^* = f(S)$$

Here, the population's average growth rate, μ, relaxes toward the equilibrium growth rate, μ^*, according to a first- or second-order model. Without going into the details of the intracellular adaptation mechanisms, this model is based on the idea that the population growth rate evolves toward that of the population if the latter is at equilibrium with the environment. The expression $\mu^* = f(S)$ corresponds to Monod's equilibrium law. The results obtained in response to a shift-up of the concentration feed (Figure 3.22) and a shift-up of the flow rate

(Figure 3.23) indicate that such a model is qualitatively relevant to describe the two types of perturbations.

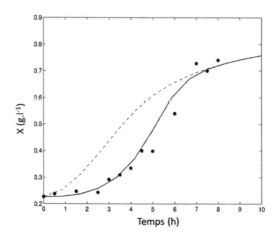

Figure 3.22. *Simulation of a chemostat's response to a sudden increase in feed concentration. Model with variable μ, second order [3.50] [MOR 09]. Data from Storer and Gaudy [STO 69]*

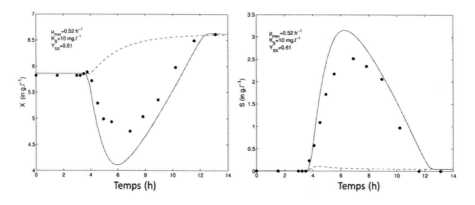

Figure 3.23. *Simulation of a chemostat's response to a sudden increase in the dilution rate. Model with variable μ, order 2 [3.50] [MOR 09]. Data from Kätterer et al. [KÄT 86]*

To conclude this section, we will mention Sweere's modeling works [SWE 88], which show us how to link the instantaneous growth rate to the metabolism. Initially, the model was based on the idea that the assimilation rate obeys an evolution law of type:

$$\frac{dq_S}{dt} = \frac{1}{\tau}\left(q_S^* - q_S\right) \ with \ q_S^* = q_{S\max}^* \frac{S}{K_n + S}$$ [3.51]

– where q_S is the instantaneous assimilation capacity;

– q_S^* is the assimilation capacity at equilibrium (balanced growth);

– τ is a characteristic time of the system regulating the adaptation of the assimilation capacity.

The objective is to take into account the inertia of the biological response by de-coupling the assimilation rate from the concentration in the liquid.

The cells cultivated in a chemostat are transferred to a closed reactor, where they receive a pulse of substrate. Concentrations are then measured every 10 min. Here, the authors conclude that the measured assimilation rate after substrate addition is lower than that observed under equilibrium conditions. In appearance, this conclusion contradicts the observations reported in section 3.3.2 and is specific to the studied yeast. However:

– the sampling frequency is low (one point every 10 min);

– this conclusion is proposed in relation to their modeling.

It is important to note that the growth rate in this model is the result of two contributions linked separately to an oxidative growth mode and a fermentative mode. Oxidative growth is limited by the availability of oxygen and respiratory capacity, whereas growth in fermentative mode exploits the possible residue of substrate left by the oxidative mode. In this modeling, the specific growth rate is algebraically related to assimilation rates:

$$\mu = Y_{S,ox}q_{S,ox} + Y_{S,red}(q_S - q_{S,ox})$$
$$q_{S,ox} = \min(q_S, Y_{OS}q_{O,\lim})$$ [3.52]
$$q_{O,\lim} = q_{O,\max} \frac{O_2}{K_O + O_2}$$

To obtain a satisfactory response in terms of cell concentration, the model must necessarily predict that the specific growth rate does not increase instantaneously but with a certain delay. The algebraic link between assimilation and growth thus requires that the assimilation rate increases in a similar way. Thus, although formulated in terms of specific assimilation rates, the proposed model uses the growth rate as the central variable. Assimilation and growth are not actually de-coupled in this model and such an approach cannot be used to assess assimilation dynamics. Indeed, the dynamic assimilation response occurs in the time interval between the first two measurement points (within the first 10 min).

Overall, it turns out that this model deals with the growth rate dynamics and it will be noted with interest that the authors identify adaptation times τ in the order of 2.5–3 h. This value is consistent with the general formula for the growth rate's dynamic response $(\tau_\mu = 1/(0.8\mu_{max})$ proposed by Morchain and Fonade [MOR 09].

3.6. Problem of coupling between a biological model and hydrodynamic model

3.6.1. *Closure of the liquid–cell transfer term*

3.6.1.1. *Algebraic relation*

In most of the models used for simulating bioreactors, the assimilation rate of the carbonaceous substrate is directly related, via a coefficient, to the growth rate and the concentration in the medium according to a Michaelis–Menten-type law [HEI 15, LAP 06, SCH 03, SWE 88]. The maximum saturation rate and constant K_S are considered to be constant. Only one type of carrier is considered. It appears that the values of these "constants", identified in a perfectly mixed homogeneous reactor, must be adjusted when we are interested in a large heterogeneous concentration bioreactor, as in the work of Vràbel [VRÀ 01]. Moreover, it is established that these constants are only apparent and vary from one study to another [KOC 82], which is logical if each experimental result is approached by a law of identical nature. In reality, it is very probable that the assimilation rate is the sum of the contributions of various carriers, each having a variable activity. This shows that some characteristics of the transfer between the cell and the culture medium are not correctly taken into account with an algebraic law. A point of improvement in the modeling of bioreactors thus lies in the prediction of the actual assimilation rates, which are not only related to the cell's growth rate but also to the activity of the various carriers.

3.6.1.2. *Expression based on the growth rate*

The studies cited above indicate that the assimilation capacity and the growth rate cannot be directly calculated from the substrate's concentration in the liquid phase. This assertion is important to consider in the case of a dynamic simulation of a bioreactor. On the contrary, for a stationary biological system (from the viewpoint of mass transfer), the cells have had the time necessary to adapt the transport systems to the substrate's availability. Under these conditions, the concentration in the medium, the specific assimilation rate and the specific growth rate are correlated. This situation must be considered as an equilibrium point of a dynamic system governed by several mechanisms: the transport of the substrate up to the membrane, the assimilation and the use of the substrate by the cell.

The first fundamental consequence stems from observations concerning the assimilation rate: if cells grown under sugar limitation are able to assimilate sugar very quickly as soon as it is given to them, then they were already able to do so but could not because of the limited supply of sugar and not because of their own (limited) assimilation capacity.

The second implication is that the modeling of the mass transfer between a cell and its environment must be re-thought. In contrast to what can be done in the case of bubble-type inclusions, drops or particles, it is not possible here to define an interfacial equilibrium law on the basis of thermodynamic quantities only. The concentration levels to which the cells were exposed modify the parameters of the assimilation law. Thus, the cell's state related to its history is present in the definition of the assimilation rate.

The third consequence is that the concept of limiting concentration is relative. The notion of limiting threshold concentration requires the knowledge of the assimilation capacity of the cells and the rate of mixing at the cellular scale. When we consider that the limitation by the substrate occurs when the characteristic time of micromixing equilibrates that of the assimilation, we have been able to establish an expression of the limiting concentration [MOR 16].

In turbulent regime, the micromixing time is given by equation [3.53], already mentioned in section 2.2.4 of Chapter 2:

$$\tau_K = 17\left(\frac{\nu}{\varepsilon}\right)^{1/2} \qquad [3.53]$$

The characteristic time of the consumption of the substrate has been established in equation [2.20] in Chapter 2. Rather than relying on the maximum consumption rate, it is possible to propose an estimate of the assimilation rate based on the real growth rate of the population, which leads to equation [3.54]. It will be recalled that the growth rate is a property of microorganisms and that it is not necessarily equal to the specific growth rate at equilibrium:

$$\tau_{R,S} = \frac{S}{Y_{SX}\,\mu X} \tag{3.54}$$

By identifying these two times, it is possible to evaluate the value of the concentration (S_{lim}), which will be perceived as limiting. Equation [3.55] leads us to express the limiting concentration as a function of the mixture's intensity, the concentration of microorganisms and their maximum assimilation capacity:

$$S_{lim} = 17\left(\frac{\nu}{\varepsilon}\right)^{1/2} Y_{SX}\,\mu X \tag{3.55}$$

Thus, it appears that the concentration from which the assimilated flux will be limited by mixing at the microscopic scale is higher when the cell density and cell demand are higher. From this, we can deduce an equation for the assimilation rate integrating the existence of a physical limitation related to the micromixing of a biological limitation by the cells' capacities:

$$q^u = \min\left(q_S^b, \frac{S}{\tau_K X}\right) \tag{3.56}$$

This expression calls for some comments:

– The quantity q_S^b refers to the assimilation capacity of the biological phase. This quantity is usually evaluated by a Michaelian law which implies an instantaneous adaptation of the transport mechanisms to the external concentration. If we use a one-equation model for the specific growth rate, we can calculate this capacity with $q_S^b = \frac{1}{Y_{SX}}\mu$.

– This law is discontinuous, which presents some risk from the numerical resolution viewpoint. Also, we proposed a continuous formulation in the form of equation [3.57], whose asymptotic behavior is similar:

$$q^u = q_S^b \left(1 - e^{-\frac{S}{S_{\lim}}} \right) \qquad [3.57]$$

– Another equivalent formulation can be established by considering that the apparent assimilation rate results from two consecutive processes, transport to the cell by the micromixing and transfer through the membrane:

$$q^u = q_S^b \frac{S}{S + S_{\lim}} \qquad [3.58]$$

It will be noted that the form is similar to a Michaelian law, where S_{lim} and q_S^b play the role of pseudo-constants, which actually vary with the culture conditions and the cells' appetite.

3.6.1.3. *Dynamic assimilation model*

We have seen that it may be necessary in a dynamic regime to de-couple the assimilation rate from the growth rate: the first one adapts quickly whereas the second one adapts to the environmental conditions over longer time periods. In the same way that a one-equation model has been established for the growth rate of the cells, a second equation can be added to describe the evolution of the substrate's assimilation capacity. On the basis of Ferenci's work [FER 99a], which indicates that the assimilation rate results from the contribution of several transport systems, a model can be proposed in the form of a sum of two contributions. The first is of non-saturable type, which introduces a notion of variable permeability of the cell's membrane P(t), and the second one is of saturable type, which refers to an active transport system:

$$q_S^b = P(t)S + q_S^{PTS} \frac{S}{k_{PTS} + S} \qquad [3.59]$$

From a general viewpoint, it may be expected that all the newly introduced quantities (in particular k_{PTS}) may depend on the state of the cell and evolve over time. In order to examine the relevancy of the approach, we will consider that only the permeability P of the membrane varies.

When the substrate concentration is high, a saturation of the assimilation capacity is observed: the second term of equation [3.59] must therefore be dominant at high concentrations. This is consistent with catabolic repression principle. When the concentration is low, the PTS system is outside its range of efficiency and the unsaturated system plays a predominant role. Natarajan and Srienc [NAT 99, NAT 00] showed that in the low substrate concentration range, the assimilation capacity of *E. coli* cells was independent of the dilution rate (hence the residual substrate concentration) . According to the proposed model, it will therefore be necessary for P(t) to increase as β decreases in order to maintain the assimilation capacity constant.

On the basis of these considerations, we propose an approach represented graphically in Figure 3.24. The biological phase is characterized by two internal variables μ^b and q_S^b. The geometric characteristics of the reactor, dimension, position of the injections, power dissipation and operating conditions (power dissipation, feed supply) and consumption by the cells define the concentration distribution in the reactor. The assimilation capacity is compared with the available flux in order to identify the value of the uptake rate q^u. This value is compared to the required rate q^r deduced from the growth potential μ^b. The gap q^d between the need and the assimilated flux serves as a measure to regulate both the growth potential over long periods of time and the assimilation capacity on shorter time scales. At equilibrium, the assimilation capacity and the assimilated flux are identical and proportional to the growth rate.

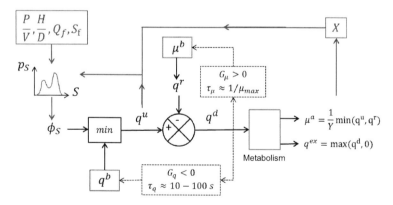

Figure 3.24. *Multi-scale model describing the hydrodynamic and biological interactions in a bioreactor. For a color version of this figure, see www.iste.co.uk/morchain/bioreactor.zip*

3.6.2. *Problem of transport and mixing*

At this stage, it appears that the description of the state of the biological phase using one or more variables was desirable or even indispensable for certain applications. We will then resort to a structured modeling by rejecting the hypothesis of equilibrium with the environment. This necessity raises a new difficulty relative to the motion of the biological phase in the physical space. The situations in which the problem of cell transport and biological phase mixing does not arise are:

– modeling of a closed reactor;

– modeling of an open reactor whose feed contains no cells.

For all other cases, such as a reactor with biomass recycling, a cascade of reactors, multi-stage reactors and *a fortiori* modeling of a reactor coupling internal hydrodynamics and biological reaction, the evolution of the properties of the biological phase should be treated carefully. Indeed, as indicated at the end of Chapter 2, the contents of the various biological particles transported by the fluid phase do not mix. Each particle preserves at the outset its properties, which evolve in response to changes in its concentration in the environment. We assume, of course, that the equilibrium with the new environment is not instantaneous.

A typical situation is illustrated in Figure 3.25. Consider the case of two chemostats fed with flow rates Q_1 and Q_2 of substrate concentrations S_{f1} and S_{f2}. It is assumed that the flow rates and feed concentrations are different so that residual substrate concentrations, cell concentrations and cell states (defined by the $p1$ or $p2$ values of the internal variable p) are different in both chemostats. If the operating conditions are kept constant for a long time (several residence times), then it is assumed that the cells are in equilibrium with the liquid phase in each reactor. It is assumed that there exists a relation to the equilibrium $\bar{p} = f(\bar{S})$ between the variables of the two phases.

Let us now examine how to describe the system's state: reactor + cells after the two currents have mixed.

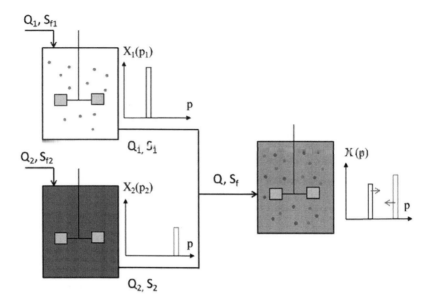

Figure 3.25. *Illustration of the problem of mixing two batches of cells each characterized by a different value of the internal property p. For a color version of this figure, see www.iste.co.uk/morchain/bioreactor.zip*

Equation [3.61] describes the overall material balance and mass conservation. Considering three reactors of identical volumes, the growth rate in the third reactor is higher than that in the first two reactors. Thus, each cell group must increase its specific growth rate. However, the amplitude of the dilution rate shift is not the same for the two groups of cells:

$$Q = Q_1 + Q_2 \qquad [3.60]$$

Equation [3.61] describes the mass conservation for the substrate. The concentration of substrate resulting from the mixing is therefore necessarily situated between S_1 and S_2. The cells are therefore also exposed to a concentration step (positive for some, negative for others):

$$QS_f = Q_1 S_1 + Q_2 S_2 \Rightarrow S_f = \frac{Q_1 S_1 + Q_2 S_2}{Q_1 + Q_2} \qquad [3.61]$$

Equation [3.62] describes the conservation of the total mass of the cells:

$$QX = Q_1 X_1 + Q_2 X_2 \Rightarrow X = \frac{Q_1 X_1 + Q_2 X_2}{Q_1 + Q_2} \qquad [3.62]$$

Obviously, the same mixing operation for the internal variable p cannot be applied because the content of the cells do not actually mix. On the contrary, if we consider that the variable p makes it possible to calculate the rate $r(p)$ of an intracellular biological process, we can calculate the average rate immediately after mixing using equation [3.63]:

$$\langle r(p) \rangle = \frac{1}{X}\left[\alpha r(p_1) + (1-\alpha) r(p_2) \right] \qquad [3.63]$$

where $\alpha = \frac{Q_1 X_1}{QX}$ is the proportion of cells from reactor 1 in the overall population after mixing. It is this proportion that defines the height of the bars on the right-hand side in Figure 3.25.

After the mixing process, each cell will evolve to adapt to the new dilution rate and concentration: a movement will take place in the property space. This movement is shown in the graph on the right-hand side in Figure 3.25 by horizontal arrows. We have seen in this chapter that we can associate an adaptation time constant to each internal variable, which is equivalent to defining a velocity in the space of internal properties. The central question therefore concerns the link between these rates (internal variables p) and the average rate of the process $\langle r(p) \rangle$.

– If the adaptation time constant for the variable p is very small compared to the residence time in the third reactor, it can be considered that the whole population instantaneously acquires a new value p_3 corresponding to the equilibrium with the concentration S_3. The knowledge of the equilibrium equation $\bar{p} = f(\bar{S})$ and the new concentration S_3 will allow the calculation of the average rate $\langle r(p) \rangle$, which will also be that of each cell. It is not strictly necessary to introduce the variable p into the facts. The equilibrium equation would be sufficient; therefore, a zero-equation model would be sufficient.

– If the adaptation time constant for the variable p is very large compared to the residence time in the third reactor, the state of each sub-population will have no time to evolve and the average rate $\langle r(p) \rangle$ will be given by equation [3.63]. Under these conditions, it is important to observe that $\langle r(p) \rangle \neq r(f(S))$ because each sub-population will remain out of equilibrium. A structured model is required to describe the initial state of each sub-population, but it is not necessary to describe the evolution of the properties.

– The intermediate situation is the most complex one. During passage into the last reactor, each cell has a sufficient time to approach the equilibrium value. Given the existence of a distribution of residence times in the reactor, the time available to each cell is different. In addition, the reactor is fed continuously by cells characterized by the value *p1* or *p2*. Thus, we will observe in the reactor a population of cells whose internal variable *p* will be distributed continuously between *p1* and *p2*. The distribution pattern, *n(p)*, is derived from the distribution of residence times and adaptation dynamics.

This last case is of course the one that requires our attention, as the first two relate to trivial situations for which answers exist. However, we are now at a dead end because structured models rely on the definition of a unique *p* value for all cells. However, the analysis of the problem showed us that it would have been wise to have a distribution function for the property *p*. In other words, it would be useful to know what fraction of the population is characterized by a certain value of *p*. In the last chapter, we will discuss the mathematical concepts and tools adapted to this problem.

Before concluding, let us observe that this information (number of individuals characterized by a given value of *p*) makes it possible to solve the problem of macromixing very naturally. A graphical illustration of this proposal is presented in Figure 3.26.

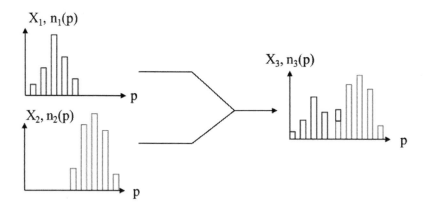

Figure 3.26. *Notion of the distribution of the variable p within the population makes it possible to manage mixing and transport problems*

The rate over the whole population (sub-divided here into NC sub-groups) is given by equation [3.64]:

$$\langle r(p) \rangle = \frac{(Q_1/Q) \sum_{j=1}^{NC} n_1(p_j) r(p_j) + (Q_2/Q) \sum_{j=1}^{NC} n_2(p_j) r(p_j)}{X} \qquad [3.64]$$

This equation can also be re-written in the form of equation [3.65], as the distribution after mixing is obtained by summing the two initial distributions weighted by a ratio of flow rates:

$$\langle r(p) \rangle = \frac{1}{X} \sum_{j=1}^{NC} r(p_j) \left[\frac{Q_1}{Q} n_1(p_j) + \frac{Q_2}{Q} n_2(p_j) \right] = \frac{1}{X} \sum_{j=1}^{NC} r(p_j) n_3(p_j) \qquad [3.65]$$

3.7. Conclusion

Biological reactors are tri-phase reactors with a solid phase in suspension. Transfer to cells and assimilation of substrates precede the transformation by metabolism. The growth rate is an integral manifestation of all these phenomena over a very long time scale compared to the transport and reaction phenomena.

The assimilation of the substrates is analogous to a transfer between the liquid and the cell. However, there is no thermodynamic law involving equilibrium between the phases. The cell modulates its assimilation capacity dynamically according to the concentrations and its needs. Adaptation dynamics are complex, involving induction times (induction of a type of carrier) and characteristic regulation times coupled with metabolism.

The characteristic times of mixing and uptake are such that the macromixing and the mesomixing compete with the assimilation. Given the turbulent nature of the flows, the cells are exposed to concentration distributions maintained by the joint and antagonistic actions of the mixing and assimilation. These local concentration distributions can be described using a transport model of the concentration variance or via population balances.

The variability between individuals within a population has two origins:

– Intrinsic (or endogenous), linked to cell multiplication. It can be represented by the inequitable distribution of mass between mother and daughter cells or by laws of probabilities relating to the parameters of the kinetic or metabolic cellular model.

– Induced (or exogenous) by changes in the environmental conditions of the cells: changes in operating conditions or heterogeneities of concentration at the local scale resulting from the competition between physical and biological phenomena.

All of these observations encourage the use of numerical tools such as population balance models to integrate aspects of micromixing and diversity within the biological phase in the mathematical description of bioreactors.

4

Biological Population Balance

4.1. Introduction

The use of the concept of population balance to describe the evolution of a population of living organisms might seem self-evident. Hatzis *et al.* [HAT 95] explain that "although continuum models have been proven adequate for many practical situations, they do not constitute the natural framework for the description of microbial population phenomena" [...] Fredrickson and co-workers [...] recognized that population models must acknowledge the segregated or corpuscular nature of microbial populations as well as the subcellular structure and composition of individual cells".

Being out of reach for a long time, information at the cellular level is now accessible through cytometry [DEL 11, DIA 10, HEW 99, LOO 05, MÜL 97, SRI 99] and single cell culture systems [KRO 08, LI 12, YAS 11]. These new experimental facts confirm that it is unlikely that all individuals of the same species are in the same physiological state (see Figure 4.1). First of all, not all individuals have the same age. We shall return to this notion, which in fact refers to the age within the cell cycle, that is, the time elapsed since the last cell division. Then, two cells with the same age do not necessarily have the same mass or the same composition, and from there, a different metabolic behavior may manifest even though they are in the same fluid medium.

It is important to realize that the great majority of experimental observations presented in the literature concern information averaged over a large number of individuals of different ages and compositions. Moreover, since the sampling frequency is generally low in the case of a conventional bioreactor culture, the collected information provides only average values over time. Finally, unless a very rapid sampling system is installed, the duration of the sampling is sufficient to ensure that cells from the entire reactor are actually collected. Thus, spatial averaging is also performed. These three dimensions, namely age, intracellular composition and position in physical space, filtered (or averaged), were used as the basis for the development of descriptive models of the microbial cultures' dynamics. From the theoretical point of view, the success of this modeling lies in the relevance of the filtering with respect to the dynamics that we are trying to describe. Since the basic experimental information was being filtered, the models were also filtered and lacked dynamics.

In addition to their conceptual appeal to deal with the case of populations, population balance frameworks force us to examine in detail the nature of the phenomena that govern the population's evolution, the characteristic times of these physical and biological phenomena, enabling us to study the dynamic interactions between a population and its environment. However, the extreme complexity of living systems, the multitude of variables required for describing them and the difficulties in solving population balance equations will be met. It will be necessary to question the origin of the diversity observed within a population: is it only the result of a redistribution of the intracellular content during the division or is it also induced by the environment's fluctuations? Is a cell able to adapt its functioning during its cycle, or is the driving force of adaptation purely a Darwinian mechanism, where the most adapted ones define the population's dynamics?

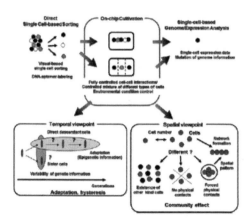

Figure 4.1. *Illustration of strategies for the retrieval of information at the cell's scale (isolated or in population) (from Yasuda [YAS 11])*

4.2. General population balance equation

4.2.1. Definitions

4.2.1.1. Physiological state vector

At the origin of any population balance model, it is necessary to distinguish individuals from each other according to a quantifiable criterion. In the case of cells, the number of criteria is absolutely gigantic if we consider:

– simple physical properties: mass, volume, diameter/length, etc.;

– the chemical composition: several thousand intracellular metabolites;

– the rate of intracellular reactions: several hundred reactions.

From a formal point of view, it is therefore possible to characterize an individual by a physiological state vector ξ comprising all or part of the above information.

$$\xi = \{\mathbf{p}, \mathbf{c}, \mathbf{v}\}$$

with

$$\mathbf{p} = \{m, v, l, d\} \qquad\qquad [4.1]$$
$$\mathbf{c} = \{c_i\}$$
$$\mathbf{v} = \{v_i\}$$

The reaction rates can be defined from the monitored intracellular concentrations and they also use additional cybernetic variables whose role is to reflect the internal regulation systems [STR 91, TAR 97, VAR 98, YOU 08]. It is not always possible to give a physical meaning to these cyber variables. In the simplest case, the cybernetic variable can be used to quantify the activity of an enzyme that catalyzes the transformation reaction of a c_1 metabolite into a c_2 metabolite.

Finally, when the notion of cell cycle is considered, the age a of the cell measured since its birth or since its entry into a particular phase of the cycle is added to the list [BIL 13, BOY 03, ROT 83, ROT 77].

The most complete case of a biological population model will thus also involve these (often intensive) variables in the definition of the state vector.

The dimension of the population balance is given by the number of elements of the state vector: $\xi = \{p,c,q,u,a\}$, where

- $p = \{m, v, l, d\}$ physical properties;

- $c = \{c_i\}$ composition;

- $q = \{q_i\}$ reaction rates;

- $u = \{u_i\}$ cybernetic variables;

- $a =$ age.

4.2.1.2. Number density function

Let us choose a value for each of these variables of the state vector and count the number of individuals in the population that actually correspond to the choice made. By dividing the number obtained, $N(\xi,t)$, by the total number of individuals in the population, we obtain the number density function $n(\xi,t)$. This function is further used to define the probability of finding a randomly chosen individual in a given physiological state ξ or probability density function, *pdf*. By construction, the total number of individuals corresponds to the integral of the number density function over all possible states,

$$\int_0^\infty N(\xi,t)d\xi = N_T(t) \qquad [4.2]$$

and the probability density function satisfies the following condition:

$$\int_0^\infty n(\xi,t)d\xi = 1 \qquad [4.3]$$

It can also be observed that by integrating this function with respect to the set of variables of the composition vectors, velocity, age and cybernetic vectors, the number density function is produced according to the physical variables

$$n(\mathbf{p},\mathbf{x},t) = \int_0^\infty n(\xi,t)dcdqda \qquad [4.4]$$

If, moreover, the only physical variable is mass, m, we define the density function in terms of the mass of individuals

$$n(m, \mathbf{x}, t) = \int_0^\infty n(\xi, t) dc dq da \qquad [4.5]$$

In the same way, we can write, for example, that

$$n(\mathbf{q}, \mathbf{x}, t) = \int_0^\infty n(\xi, t) dp dc da \qquad [4.6]$$

Finally, we note that the number density function, of let us say cell age α, can be written as the ratio of the total number of particles of age α regardless of their mass, composition, etc., and the total number of particles of all ages, all masses, all compositions, etc.,

$$n(\alpha, t) = \frac{\int_0^\infty N(\mathbf{p}, \mathbf{c}, \mathbf{q}, \alpha, t) dp dc dq}{\int_0^\infty N(\xi, t) dp dc dq da} \qquad [4.7]$$

4.2.1.3. Average value

The average value of any variable at the population scale[1] is deduced from the moments m_k, of the number density function.

$$M_k = \int_0^\infty v^k n(v, t) dv \qquad [4.8]$$

1 The notation ~ means that the average is taken on the biological phase.

Thus, the population average value is defined as the ratio of the first-order moment to the zero-order moment,

$$\tilde{v} = \frac{M_1}{M_0} = \frac{\int_0^{\infty} v^1 n(v,t)dv}{\int_0^{\infty} v^0 n(v,t)dv} = \frac{1}{N_T}\int_0^{\infty} vn(v,t)dv \qquad [4.9]$$

Thus, fundamental equation [4.10], integrated with respect to all variables, must naturally lead to the equation we write when considering a structured model based on the average individual.

Figure 4.2, adapted from the book by Bailey and Ollis, illustrates this point [BAI 86]. The distinction between segregated and non-segregated models is quite clear and the transition from one to the other is a mathematically rigorous averaging operation. The passage from structured to unstructured is much less clear. In practice, this passage is never performed because the considered internal variables are limited in number and do not describe the entire intracellular mass. The exercise therefore remains rather intellectual.

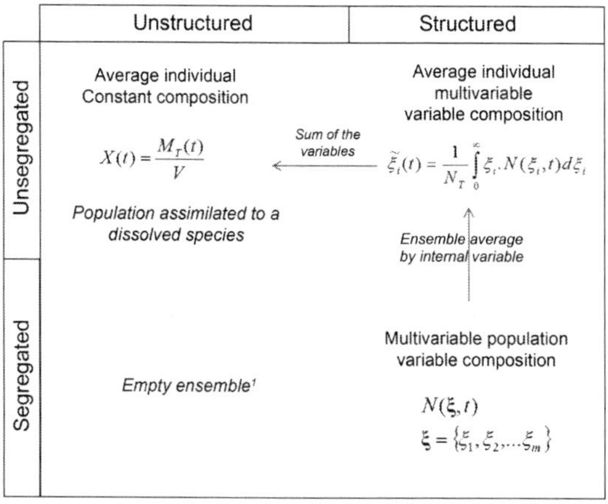

	Unstructured	Structured
Unsegregated	Average individual Constant composition $X(t) = \dfrac{M_T(t)}{V}$ Population assimilated to a dissolved species	Average individual multivariable variable composition *Sum of the variables* ← $\tilde{\xi}_i(t) = \dfrac{1}{N_T}\int_0^{\infty} \xi_i . N(\xi_i,t)d\xi_i$ *Ensemble average by internal variable*
Segregated	*Empty ensemble¹*	Multivariable population variable composition $N(\xi,t)$ $\xi = \{\xi_1, \xi_2, \dots \xi_m\}$

Figure 4.2. *Classification of biological phase description models in relation to modeling of biological reactions. A segregated model distinguishes between individuals. In the strict sense, it is therefore structured. Sometimes, however, models that distinguish individuals according to a physical property which is not related to reaction kinetics, are placed here (according to Bailey and Ollis [BAI 86])*

4.2.1.4. *Establishment of the fundamental equation*

The general formulation of the population balance equation as it is used today in the field of process engineering (but not only) is due to Ramkrishna's work [RAM 71]. It is based on the equation proposed by Schmoluschovski in 1912. This equation describes the evolution of a density function. It is a function that informs about the probability of finding a particle in a particular state at a point in space at a given time. As in the case of a concentration of species[2] in solution, a conservation equation can be established for such a function,

$$\frac{\partial n(\xi,x,t)}{\partial t} + \frac{\partial}{\partial \xi}\left(n(\xi,t)u_{\xi}\right) + U_{i}\frac{\partial n(\xi,x,t)}{\partial x_{i}} - \frac{\partial}{\partial x_{i}}\left(D\frac{\partial n(\xi,x,t)}{\partial x_{i}}\right) = S(\xi,t)$$

[4.10]

where

$- n(\xi)$: number density function;

$- n(\xi)d\xi$: number of individuals whose property vector is between ξ and $\xi + d\xi$;

$- x$: vector defining the position in the physical space (x, y, z);

$- U_{i}$: transport velocity in the direction i of the physical space;

$- D$: diffusion coefficient (diffusive transport in the physical space);

$- u_{\xi}$: transport velocity in the space of internal properties;

$- S(\xi,t)$: source term representing the net production rate of individuals characterized by the property vector ξ.

This equation may seem rather complex but it describes several phenomena that are in reality quite simple. Here, we will illustrate the meaning of the different terms.

Let us begin by writing the equation of conservation of mass of a compound in a fluid volume of dimension *dxdydz*, fixed in space. C is the compound's concentration in $kg.m^{-3}$. First, a purely convective transport is considered in the absence of a chemical reaction.

---

2 The concentration is a (number) density function: the number of molecules per unit volume.

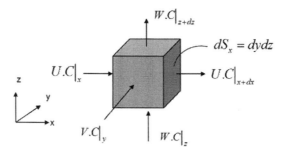

Figure 4.3. *Diagram illustrating the fluxes entering and leaving an elementary volume*

The mass variation of C contained in the volume results from the sum of the fluxes entering and exiting all of the facets that define the contour of the considered volume (see Figure 4.3). It will be noted here that U, V and W are the components of the fluid's velocity in the directions x, y and z, respectively. The flux entering the facet $dydz$ at the abscissa x is written as $\Phi(x) = U.C|_x dydz$, the one exiting the facet of the same surface situated at $x + dx$ is written in the same way as $\Phi(x + dx) = -U.C|_{x+dx} dydz$. By convention, the plus sign is used for an entering flux and the minus sign for a flux exiting the fluid domain. Using the theorem of finite increments, we observe that the difference between these fluxes can be written as a derivative of the flux in space,

$$U.C|_x\, dydz - U.C|_{x+dx}\, dydz = -\frac{\partial (U.C)}{\partial x} dxdydz \qquad [4.11]$$

The same process is carried out in the three directions of space to obtain equation [4.12],

$$\frac{\partial Cdv}{\partial t} = -\frac{\partial (U.C)}{\partial x}\underbrace{dx.dS_x}_{dv} - \frac{\partial (V.C)}{\partial y}\underbrace{dy.dS_y}_{dv} - \frac{\partial (W.C)}{\partial z}\underbrace{dz.dS_z}_{dv} \qquad [4.12]$$

From a more general point of view, it is not required for the volume's facets to be aligned with the coordinate system. In this case, the sum of incoming and outgoing flows is written as the integral on this contour of the volume of the scalar product between the velocity vector at a point of the contour and the vector normal to the surface at this point. Moreover, the

Green–Ostrogradski theorem makes it possible to convert the integral of the flux on the contour into an integral of the divergence of the vector field on a volume. Thus, we can see equation [4.13] as a generalization of equation [4.11],

$$\iint_{\Omega} \vec{U}.\vec{n}.C dS = \iiint_{V} div(\vec{U}C)dv \qquad [4.13]$$

Let us now take equation [4.12] and decompose the divergence term,

$$\frac{\partial C}{\partial t} = -div(\vec{U}.C) = -Cdiv(\vec{U}) + \vec{U}.\overrightarrow{grad}(C) \qquad [4.14]$$

However, in an incompressible flow, the divergence of the velocity field is zero, which ultimately leads to equation [4.15] that we will write using different equivalent formalisms as follows:

$$\frac{\partial C}{\partial t} + \vec{U}.\overrightarrow{grad}(C) = 0$$

$$\frac{\partial C}{\partial t} + U_i \frac{\partial C}{\partial x_i} = 0 \qquad [4.15]$$

$$\frac{\partial C}{\partial t} + \vec{U}\vec{\nabla}C = 0$$

It is sufficient to take into account the possibility of a scalar transport by diffusion [4.16] and the presence of a source term (net production related to transfers and/or chemical reactions) to obtain the general equation for the conservation of mass of compound C [4.17],

$$\Phi = U.C - D\nabla C \qquad [4.16]$$

$$\frac{\partial C}{\partial t} + U.\nabla C - \nabla(D\nabla C) = S_C \qquad [4.17]$$

The final step is to broaden our appreciation of the notions of space and concentration. It is generally understood that space consists of three dimensions so that C implicitly evokes C(x, y, z). Moreover, the concentration echoes a number of particles of matter of the same nature called molecules[3]. Let us now imagine that each particle may be characterized by one or more properties ξ. We will then have to deal with a new space of properties in addition to physical space. In the same way that a particle's velocity in physical space corresponds to the variation in its position (x_p, y_p, z_p) over time, the velocity in the property space will be defined as the temporal variation in the value of a particle's property,

$$u_{\xi_i} = \frac{\partial \xi_i}{\partial t}$$ [4.18]

For example, if we are interested in droplets and in the evolution of their mass, it will be observed that a droplet that exchanges matter has a velocity in the mass space. Figure 4.4 illustrates this notion: in each direction of the property space, we can express the conservation of the number of particles in a given state (value of property i comprised between ξ_i and $\xi_i + d\xi_i$) as the difference between the incoming fluxes at ξ_i and outgoing fluxes at $\xi_i + d\xi_i$.

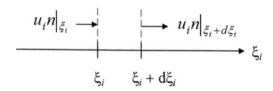

Figure 4.4. *Illustration of transport in the property space*

The conservation equation for the number of particles in a given state can be seen as a "classical" conservation equation supplemented with a term accounting for the transport of particles in the particles properties' space. We shall note that the divergence of the velocity in the property space is not zero, which explains the formulation difference between the two transport terms in equation [4.10].

The real peculiarity of a population balance equation lies in the second part that we have not explained so far. Let us return to the example of our droplets. Consider a group of droplets having a particular mass. Under the effect of breakage and coalescence, droplets will join or leave this group:

---

3 We could have reasoned in moles in order to establish the conservation equation.

– any droplet of mass m which undergoes a breakage leaves the group;

– any droplet of mass m which merges with any other droplet leaves the group;

– a fraction of the droplets of mass $m' > m$, undergoing breakage, will produce a droplet of mass m;

– any coalescence of two droplets of masses $m-m'$ and m' will produce a droplet of mass m.

The formulation of the second part in this case will involve several integrals [KUM 96a],

$$S_C = -\Gamma(m)n(m) - \int_0^\infty n(m)n(m')Q(m,m')dm' + \ldots$$

[4.19]

$$\ldots \int_m^\infty p(m,m')n(m')\Gamma(m')dm' + \int_0^m n(m')n(m'-m)Q(m',m'-m)dm'$$

– Q is the aggregation (coalescence) frequency;

– Γ is the breakage frequency;

– $p(m,m')dm'$ is the number of particles of mass m formed following the rupture of particles of size m'.

The identification of these laws on the basis of theoretical considerations and experimental data is a central issue in the field of population balances. This often implies an inverse problem: how to identify elementary laws from the observation of the consequences of their interactions on a population's characteristics [BIL 13, LEB 74].

The second issue concerns the resolution of these so-called integro-differential equations because they associate integrals and partial derivatives. The scientific advancement in this field is impressive, to the extent of the difficulty of the problem. Advanced skills in mathematics are mandatory to search for analytical solutions. There exist a number of them for particular problems. Resolution by various numerical methods is not basically easier. The complexity increases strongly with the number of properties chosen to describe the state of the particles. Reference may be made to the work of Rotenberg [ROT 77, ROT 83, KUM 96a].

4.2.2. *What are the links between structured models and population balances?*

Structured models use, by definition, intracellular variables, without distinguishing between individuals, hence we have a model based on an average individual. We shall see here what is covered by such a model. The equation describing the average population dynamics can, of course, be expected to be similar to the equation describing the dynamics of the average individual. However, similar does not mean identical and we will show the necessary conditions for equivalence between these two approaches. We rely here on the work of Mantzaris [MAN 05, MAN 06], enriched with personal comments.

4.2.2.1. *Demonstration*

For a given intracellular compound, c, the equation of conservation of mass in an open reactor of constant volume V is written as:

$$\frac{dVXc}{dt} = VX\frac{dc}{dt} + Vc\frac{dX}{dt} \cdots \Rightarrow \frac{dc}{dt} = R(c) - \mu.c - \frac{Q}{V}c \qquad [4.20]$$

$- X$ denotes the cell concentration (in g.l^{-1});

$- R(C)$ represents the net production term of c linked to the set of intracellular reactions;

$- \mu$ is the specific growth rate (h^{-1}).

Thus, the second term on the right-hand side is generally described as a term for the mass dilution of c contained in the cells because of their multiplication [NIE 92]. This denomination can be understood as follows: in the absence of production, the initial mass of c decreases exponentially as the cells multiply. At the level of an average cell, the division dilutes the intracellular compounds.

Consider now the population balance equation based on the same variable c.

$$\frac{\partial N(c,t)}{\partial t} + \frac{\partial}{\partial c}\left[r(c)N(c,t)\right] = \cdots$$

$$2\int_{m}^{\infty}\Gamma(c)p(c,c')N(c',t)dc' - \Gamma(c)N(c,t) - \frac{Q}{V}N(c,t) \qquad [4.21]$$

Multiply this equation by c and integrate over $[0, + \infty]$

$$\int \frac{\partial c N(c,t)}{\partial t} dc + \int c \frac{\partial}{\partial c} [r(c) N(c,t)] dc = ...$$

$$2 \int c \int_{c}^{\infty} \Gamma(c) p(c,c') N(c',t) dc' dc - \int c \Gamma(c) N(c,t) dc - \frac{Q}{V} \int c N(c,t) dc$$
[4.22]

Let us look at each of the terms of this equation: by using the definition of the average value of c given in equation [4.9], the first term on the left-hand side becomes:

$$\int \frac{\partial c . N(c,t)}{\partial t} dc = \frac{\partial \int c . N(c,t) dc}{\partial t} = \frac{d(\tilde{c} N_t)}{dt} = \tilde{c} \frac{d(N_t)}{dt} + N_t \frac{d\tilde{c}}{dt}$$
[4.23]

Then, by integrating by part the second term on the left-hand side, we obtain

$$\int c \frac{\partial}{\partial c} [r(c) N(c,t)] dc = \underbrace{[c . r(c) N(c,t)]_{0}^{\infty}}_{0} - \int r(c) N(c,t) dc$$
[4.24]

and by conservation of the total mass of c during the cell division process, we get

$$\int_{0}^{\infty} \left[c \int_{c}^{\infty} \Gamma(c') p(c,c') N(c',t) dc' - c \Gamma(c) N(c,t) \right] dc = 0$$
[4.25]

We finally obtain the equation describing the evolution of the average of c

$$N_t \frac{d\tilde{c}}{dt} = \int r(c) N(c,t) dc - \tilde{c} \frac{dN_t}{dt} - \frac{Q}{V} \tilde{c} N_t$$
[4.26]

We can then divide this equation by the total number of cells N_t and observe that, by definition of the specific growth rate, we have:

$$\tilde{\mu} = \frac{1}{N_t} \frac{dN_t}{dt}$$
[4.27]

Hence, the following final equation for the average of c, similar enough to the equation of the continuous structured model of equation [4.20] recalled below:

$$\frac{d\tilde{c}}{dt} = \int r(c)n(c,t)dc - \tilde{\mu}.\tilde{c} - \frac{Q}{V}\tilde{c}$$

[4.28]

$$\frac{dC}{dt} = R(C) - \mu C - \frac{Q}{V}C$$

NOTE.– In a structured model, intracellular biological variables represent[4] an average value at the population scale $C \equiv \tilde{c}$. Accordingly, $R(C)$ now clearly appears as a *model* for the integral of intracellular reactions based on the average value, C, of the intracellular compound. Also, μ stands for the population's mean specific growth rate.

4.2.2.2. Equivalence conditions

The two expressions of equation [4.28] are strictly equivalent if:

– condition C1: trivially, all individuals have the same composition;

– condition C2: the reaction rate r is linear in c.

Then we have:

$$\int r(c)n(c,t)dc = k\int cn(c,t)dc = k\tilde{c}$$

[4.29]

It is rather interesting to study the cases of reaction rates, which obey a fundamental Michaelis–Menten-type biological law. We then have:

$$r(c) = r_{max}\frac{c}{k_c + c}$$

[4.30]

This rate law is quasi-bilinear: proportional to c at low values ($c \ll k_c$) and constant at high values ($c \gg k_c$). Thus, condition C2 is in practice verified in many practical situations. Therefore, the non-segregated

---

4 It is indeed an equivalence and not an equality.

structured model will be as relevant as a population balance model in describing the average population dynamics.

Conversely, if the value of c fluctuates around k_c, sometimes below and sometimes above, then the continuous structured model will no longer be equivalent to the population balance model, as its ability to describe the average dynamics will be degraded. It is typically the situation that occurs in a heterogeneous reactor: microorganisms are alternately subjected to high and low concentrations.

In the same way, if a metabolic model is used, the population-based approach is preferable because it allows us to consider the reaction rates for each individual rather than the averaged reaction rate over all individuals.

4.3. Illustrative examples

4.3.1. *Models based on mass discrimination*

Among the earliest studies of the population balance of cell populations are the contributions of Fredrickson and Tsuchiya [FRE 63] and Subramanian *et al.* [SUB 70]. Discrimination between individuals is based on their mass m. The basic equation used is that proposed by Eakman and leads to the following balance in the case of a perfectly stirred reactor of volume V, supplied in continuous mode at flow rate Q,

$$\frac{\partial n(m,t)}{\partial t} + \frac{\partial}{\partial m}\left[r(m,c_s)n(m,t)\right] = \dots$$

$$\dots = 2\int_m^\infty \Gamma(m',c_s)p(m,m')n(m',t)dm' - \Gamma(m,c_s)n(m,t) - \frac{Q}{V}n(m,t)$$

[4.31]

where:

– $n(m,t)dm$: number of cells per unit volume with a mass between m and $m + dm$ at time t;

– $r(m,c_s)$: growth rate or mass growth rate of a cell of mass m in a medium of concentration c_S [g.s^{-1}];

– $\Gamma(m,c_S)$: breakage frequency of a cell of mass m in a medium of concentration c_S;

$-p(m,m')$: probability of forming a daughter cell of mass m from a mother cell of mass m'.

Assuming that the mass distribution of the cells produced during cell division obeys a Gaussian law centered on the average mass during division, m_c, Eakman proposes a rupture law in the form:

$$\Gamma(m,t) = \frac{2}{\sigma\sqrt{\pi}} \frac{e^{-\left(\frac{m-m_c}{\sigma}\right)^2}}{erfc\left(\frac{m-m_c}{\sigma}\right)} r(m,c_s)$$ [4.32]

The variations in liquid-phase concentrations induced by biological production and consumption are described as follows:

$$\frac{dc_s}{dt} = \underbrace{\frac{Q}{V}\left(c_{s,f} - c_s\right)}_{\text{Convective term}}$$

$$\underbrace{-\int_0^\infty \left[\beta(m)r_1(m,c_s) - \gamma(m)r_2(m)\right] n(m,t)dm}_{\text{source terms}}$$ [4.33]

$-\beta(m)$: conversion yield of the assimilated substrate into biomass for a cell of mass m;

$-\gamma(m)$: conversion yield of matter released into the substrate for a cell of mass m;

$-r_1(m,c_s)$: assimilation rate of a cell of mass m in a medium of concentration c_S [g.s^{-1}];

$-r_2(m)$: mass release rate of a cell of mass m [g.s^{-1}].

In the original work, the integral's lower bound is denoted by m and not 0. Now, we must consider the transfers for all cells whatever their size. However, if we assume that every daughter cell from the division has a minimum size, m_{min}, then we can limit ourselves to integrating between m_{min} and infinity. Here, we choose to integrate over the entire interval while observing that the function $n(m,t)$ tends to 0 at the bounds.

Definitions of assimilation and release rates are derived from a reasoning attributed to Von Bertalanffy [VON 42] specifying that the assimilation rate is proportional to the cell's surface while the release rate is related to cell's activity, which is proportional to the mass and to the maximum value of the specific growth rate, μ_{max}, which is supposed to be known:

$$r_1(m,c_s) = \phi_{max} \underbrace{\left(\frac{m}{\rho d}\right)}_{Cellsurface}\left(\frac{c_s}{k_s + c_s}\right)$$

[4.34]

$$r_2(m) = \mu_{max} m$$

– ϕ_{max}: maximum mass flux linked to the assimilation of the substrate through the cell wall.

This formulation clearly identifies the culture medium to a system consisting of several phases since it posits that the variations in the liquid-phase concentration are the consequence of the mass transfers between the liquid and the cells in suspension.

4.3.1.1. Resolution

The solution to the problem is obtained by the weighted residual method. The details go beyond the scope of this book. In a few words, it is a search for a solution in the form of a series of weighted-type function. Here, the type functions are Laguerre polynomials and the chosen weighting functions are $x^n e^{-\lambda x}$ with $\lambda > 0$,

$$r_1(m,c_s) = \varphi_{max} \underbrace{\left(\frac{m}{\rho d}\right)}_{Cell\ surface}\left(\frac{c_s}{k_s + c_s}\right)$$

[4.35]

$$r_2(m) = \mu_{max} m$$

By introducing this form of solution into the initial equation, we form a system of equations which relates to the coefficients λ_i of the weighting functions. The solution is obtained by minimizing the residuals for all the considered values of m.

4.3.1.2. Significant results

Figure 4.5, from Subramanian *et al.* [SUB 70], shows the evolution of the cell mass distribution over time during a batch culture. The initial conditions used are given by the set of equations [4.36].

$$n(m,0) = n_0 \frac{m}{\varepsilon} e^{-\frac{m}{\varepsilon}}$$

$$\varepsilon = \frac{\sqrt{2}}{10} m_c \qquad\qquad [4.36]$$

$$C(0) = \int mn(m,0)dm = 0.36 g.l^{-1}$$

$$c_S(0) = 2 g.l^{-1}$$

The number next to each curve indicates the time in hours. The initial distribution consists of small cells whose mass increases up to 0.8 hours. From this moment, the mass distribution no longer changes, only the total number of cells increases. This balanced growth phase lasts until t = 1.2 hours after which the depletion of the substrate leads to the cessation of cell multiplication. The distribution then slides to the right (t = 1.4 hours) because the release rate is not zero.

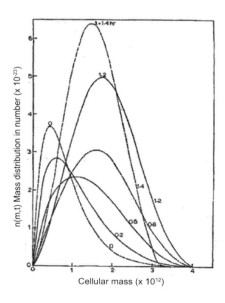

Figure 4.5. *Variation in the mass distribution in batch culture [SUB 70]*

Figure 4.6 shows the evolution of the substrate and biomass concentration as well as the total number of cells in batch culture. The number of cells stabilizes after the depletion of the substrate (t = 1.2 hours) but the total mass of the cells decreases thereafter. This is due to the fact that there is no death term and that the release rate depends only on the mass of the cells.

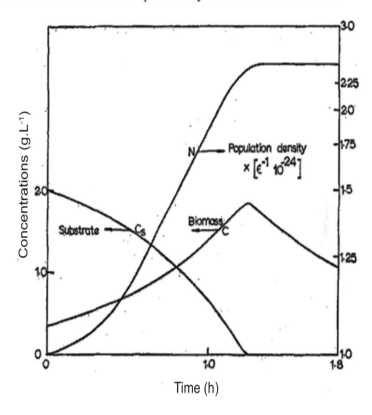

Figure 4.6. *Evolution of the total number of individuals N, the associated mass C and the substrate concentration CS in a batch culture [SUB 70]*

Figure 4.7 shows the evolution of the substrate and biomass concentration as well as the total number of cells in a chemostat. The initial conditions are chosen so that the mass distribution becomes rapidly stationary. It is then observed that the total mass and the total number of cells increase in parallel: this is the famous balanced growth phase.

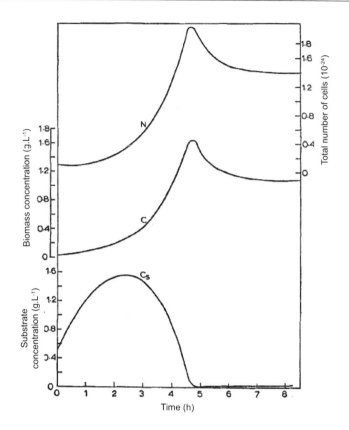

Figure 4.7. *Evolution of the total number of individuals N,
the associated mass C and the substrate concentration CS
in a continuous culture [SUB 70]*

We shall show in the next section that the evolution of the total mass
obeys an exponential law,

$$\frac{dM_1}{dt} = \tilde{\mu} M_1 \tag{4.37}$$

From the moment when the distribution $n(m,t)$ is stationary, the average
mass is constant. As the authors use a number density function in mass, this
average corresponds to the ratio of the moment of order 1 (total mass) over
the moment of order 0 of the distribution (total number), hence we obtain
equation [4.38],

$$\tilde{m} = \frac{M_1}{M_0} \tag{4.38}$$

It can be deduced that, from the moment when the distribution is stationary, the total number of particles evolve exponentially at the same rate as the total mass.

At the beginning of the culture, the curves showing the evolution of the total number of cells, like that of the total biomass, Figures 4.6 and 4.7, provoke the appearance of a behavior that could be interpreted as a latency phase analogous to what is experimentally observed. However, the phenomenon is related in this case to the choice of the initial distribution. Regardless of concentration, cells with a high mass grow rapidly and those with a low mass grow slowly. Thus, if we start with a population of high-mass cells in a rich medium, the model will not predict a latency phase. On the other hand, if we start with cells with a low mass, we will observe a phase of prolonged latency under the same conditions. Let us now imagine cells of the same mass belonging to two populations of which one is cultured in batch and the other in low dilution ratio chemostat. Some are accustomed to a high sugar concentration and others to a residual concentration lower than K_S. Placed in the same environment with a high substrate concentration, all the cells, because they have the same mass, will instantly adopt the same growth rate, according to law [4.34]. However, the change in environment will be virtually zero for the cells coming from the batch but very important for those coming from the chemostat.

Thus, the authors report that the use of a single global variable such as mass (or age) is probably insufficient to describe the physiological state of a particular cell if we intend to account for the fact that the cell's response to its environment depends on the cell's origin (its history in this sense). A description of the state using an internal variable vector would probably be preferable, but the size of this vector should be kept as small as possible in order not to risk making mathematical developments unmanageable.

NOTE.– All of the problem's dimensions are already posed: population heterogeneity, definition of a cellular state allowing to describe the reactions as well as adaptation to the environment, functioning of the microorganisms based on the exchanges between phases.

4.3.1.3. *Limits of the mass-based approach*

We shall show here that, if the mass growth rate m is a linear function of mass, we will observe an exponential population growth in mass. In doing so, we will also show that this function must be written $\dot{m} = \tilde{\mu}(S).m$, and we will discuss the contradiction of using an integrated information at the population scale, that is, $\tilde{\mu}(S)$, to describe the behavior of a fraction of the population: individuals of mass m. Then, we will establish the general condition to observe an exponential growth. We will show that this condition relates to the existence of a stationary distribution rather than to the formulation of the velocity $\dot{m}$.

Let us start with the equation of the probability density function in the number of the cells' mass $n(m,t)$ and calculate its first moment, namely the cells' total mass,

$$\frac{\partial n(m,t)}{\partial t} + \frac{\partial}{\partial m}\left[\dot{m}n(m,t)\right] = 2\int_{m}^{\infty}\Gamma(m')p(m,m')n(m',t)dm' \\ -\Gamma(m)n(m,t) \qquad [4.39]$$

Let us leave aside the time dependence to lighten the writing and multiply equation [4.39] by m and integrate into the domain $[0, +\infty]$, assuming that there is no cell with a negative mass,

$$\int m\frac{\partial n(m)}{\partial t}dm + \int m\frac{\partial}{\partial m}\left[\dot{m}n(m)\right]dm = ... \\ ... = 2\int m\int_{m}^{\infty}\Gamma(m)p(m,m')n(m')dm'dm - \int m\Gamma(m)n(m,t)dm \qquad [4.40]$$

Since variables m and t are independent[5] [SCO 68, MCG 97, MAR 03, MAN 06, MAN 07], we get equation [4.41],

$$\int m\frac{\partial n(m)}{\partial t}dm = \frac{d}{dt}\int mn(m)dm \qquad [4.41]$$

The total cell mass is defined by equation [4.42]

---

5 The mass of a cell varies over time.

$$M_1 = \int_0^\infty mn(m)dm \tag{4.42}$$

Hence, we deduce that the first term of the left-hand side of equation [4.40] describes the evolution of the cells' total mass,

$$\int m \frac{\partial n(m)}{\partial t} dm = \frac{dM_1}{dt} \tag{4.43}$$

The second term of the left-hand side of equation [4.40] is integrated in parts and is simplified by considering that the fluxes at the bounds are necessarily zero, otherwise we would produce cells of negative or infinite mass,

$$\int m \frac{\partial}{\partial m}[\dot{m}n(m)]dm = \underbrace{[m.\dot{m}n(m)]_0^\infty}_{0} - \int \dot{m}.n(m)dm \tag{4.44}$$

Conservation of the cells' total mass during the division process requires that the integral of the term on the right-hand side of equation [4.40] is globally zero.

$$\int_0^\infty \left[m \int_m^\infty \Gamma(m')p(m,m')n(m')dm' - m\Gamma(m)n(m) \right] dm = 0 \tag{4.45}$$

By introducing equations [4.43] and [4.44] into equation [4.40], we obtain equation [4.46]

$$\frac{dM_1}{dt} = \int \dot{m}n(m,t)dm \tag{4.46}$$

Let us now suppose that the mass growth law is a linear function of the mass, $\dot{m} = \alpha.m$. We shall obtain the well-known result according to which the mass evolves exponentially in the absence of limitation,

$$\frac{dM_1}{dt} = \int \alpha mn(m)dm = \alpha M_1 \tag{4.47}$$

The fact that this general behavior of the population is reproduced is wrongly interpreted as a validation of the law of growth. Indeed, we must conclude from equation [4.47] that the coefficient we have just introduced is nothing more than the population's specific growth rate, that of the whole population, and not just that of the individuals of mass m.

Thus, we have just demonstrated that in order to obtain an exponential growth of the population at the specific rate $\tilde{\mu}$, it has been assumed that each cell, whatever its mass, has the same specific growth rate as the population as a whole. This is obviously a tautology that discredits the whole approach because we use a macroscopic law to describe the behavior of a subgroup of individuals. In a report by the European Molecular Biology Organization, Boye and Nordström use a purely microbiological argument based on experimental observations to confirm that mass is certainly not the relevant parameter for coupling growth and cell cycle [BOY 03].

Formulation [4.48] is more complete and apparently more relevant because it considers the substrate concentration in the medium. However, it suffers from the same default since it is still a linear function of the mass. It is easy to show that we will then have $\alpha = \mu_{max}$. Moreover, this law postultes an immediate adaptation of the indivual's growth rate to the concentration variations in the liquid phase.

$$\dot{m} = \alpha m \frac{S}{K_S + S} \qquad [4.48]$$

Let us rewrite the initial equation [4.40] in light of these observations,

$$\frac{\partial n(m)}{\partial t} + \frac{\partial}{\partial c}\left[m.\mu_{max} \frac{S}{K_S + S} .n(m) \right] = 2\int_m^\infty \Gamma(m')p(m,m')n(m')dm' \qquad [4.49]$$
$$-\Gamma(m)n(m)$$

It is therefore clear that the population's specific growth rate is not the result of integration over all individuals but that the average value is imposed on each individual. Moreover, each individual is supposed to develop at the equilibrium rate (in the sense used in Chapter 3), which removes any dynamic character from the modeling. The distribution $n(m)$ only evolves because the initial distribution is not the equilibrium distribution. Finally, it is observed that, since the relationship between mass growth rate and mass is linear, it is not necessary to develop a population balance since the population's average rate is calculated directly from the individuals' average mass.

Let us now turn to the calculation of the evolution of the total number of cells, that is to say the zero-order moment of distribution $n(m)$. Let us integrate the equation for all masses between 0 and $+\infty$,

$$\int_0^\infty \frac{\partial n(m)}{\partial t}\,dm + \int_0^\infty \frac{\partial}{\partial c}[r(m)n(m)]\,dm = \int_0^\infty \int_m^\infty 2\Gamma(m')p(m,m')n(m')\,dm'dm$$

$$-\int_0^\infty \Gamma(m)n(m)\,dm \qquad [4.50]$$

The first term on the left-hand side corresponds to the temporal derivative of the total number of cells, whereas the second term on the left-hand side is null because the fluxes at 0 and $+\infty$ are zero. The first term on the right-hand side remains yet to be studied,

$$\int_0^\infty \int_m^\infty 2\Gamma(m')p(m,m')n(m')\,dm'dm = \int_0^{m'} \int_0^\infty 2\Gamma(m')p(m,m')n(m')\,dm'dm \qquad [4.51]$$

By rearranging this integral, it is established that

$$\int_0^{m'} \int_0^\infty 2\Gamma(m')p(m,m')n(m')\,dmdm' = 2\int_0^\infty \Gamma(m')n(m')\left[\int_0^{m'} p(m,m')\,dm\right]dm' \qquad [4.52]$$

The value of the integral between brackets is 1 because $p(m,m')$ is the probability of obtaining a cell of mass m during the rupture of a cell of mass m'. Equation [4.58] of the evolution of the zero-order moment is finally obtained.

$$\frac{\partial M_0}{\partial t} = \int_0^\infty \Gamma(m)n(m,t)\,dm \qquad [4.53]$$

In the balanced growth phase, the total mass and the total number evolve in parallel, so there is proportionality between the temporal derivatives of the moments of orders 1 and 0, which leads to the expression linking the mass growth rate and division frequency. It is sufficient that the mass distribution is stationary and it is the cell number proliferation which leads to an exponential mass growth.

NOTE.– It is sufficient that the mass distribution is stationary to observe an exponential growth in number. Since the average mass is constant the cell number proliferation leads to an exponential mass growth. This result is not dependent on the exact formulation of the velocity $\dot{m}$

$$\int_0^\infty \Gamma(m)n(m,t)dm \approx \int_0^\infty r(m)n(m,t)dm \qquad [4.54]$$

Before proceeding, let us return to the expression used in equation [4.34] to define the relationship between the assimilation rate and the cell's surface. To reach this expression of a cell's surface $S = m/\rho d$, a hypothesis on the form has been made. Von Bertalanffy considered the case of cylindrical cells of constant diameter, d, and density, ρ, whose mass gain results in an elongation. His proposal is relevant for a bacterium such as *Escherichia coli*. The visualizations presented in the recent work of Nobs and Maerkl [NOB 14] indicate that this hypothesis is well suited for the yeast *Schizosaccharomyces pombe*. The process of cell multiplication is essentially explained by an elongation phase of the cell followed by a division.

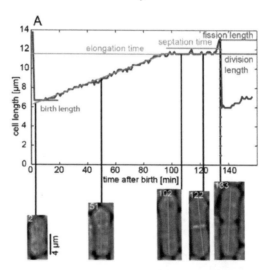

Figure 4.8. *The length of S. pombe cells evolves linearly with time during the division cycle [NOB 14]. For a color version of this figure, see www.iste.co.uk/morchain/bioreactor.zip*

It is observed that the elongation phase is a linear function of time. If we retain the hypothesis of a constant cell density ρ, and a cylindrical form, we deduce two equations, [4.55] and [4.56].

$$A \approx \pi l d \approx \pi(\alpha t)d \qquad\qquad [4.55]$$

$$m = \rho v \approx \rho \pi(\alpha t)d^2 \Rightarrow \dot{m} \approx \pi \rho \alpha d^2 = k \qquad\qquad [4.56]$$

The mass approach leads to several roadblocks at the level of the formulation of the growth law. Moreover, knowledge of a particle's mass does not give any direct information on the rates of the reactions involved. It seems, therefore, that in order to approach the problem, it is necessary to have a length scale (at constant ρ this may correlated with the cell mass) and a time scale such as the cell cycle's duration or a synthesis rate of the intracellular products.

4.3.2. Differentiation by age

4.3.2.1. Equation formation

At first, the formalism is identical to a classical population balance. The concept is improved by introducing the cell cycle notion [FRE 63, LEB 74, ROT 83]. The principle is to distinguish individuals according to two (or more) characteristics, usually mass and age. The transition between the different phases of the cell cycle is based on the cells' age while mass variations (and therefore exchanges with the fluid medium) obey laws specific to each phase of the cell cycle. Moreover, cell multiplication is described by a breakage mechanism.

In theory, it is therefore a multi-dimensional population balance. A cell is characterized by its mass, age and the phase of the cycle in which it is found. However, we will see that Hatzis *et al.* will initially dissociate the evolution of age from mass [HAT 95]. The authors set out the principles of this approach by discussing, in particular, the nature of the transition functions between the phases of the cell cycle:

– the transition is instantaneous when the age (or mass) within a phase reaches a critical value. This simple proposition (zero or infinite transition rate) is however not realistic and does not agree with the observations. Its discontinuous character induces numerical problems;

– the transition speed is constant beyond a limit value of age (or mass). It is then shown that this leads to a breakage law of the type

$$
\begin{cases}
\Gamma(x) = \gamma e^{-\gamma(x-x_c)} & x \geq x_c \\
\Gamma(x) = 0 & \text{otherwise}
\end{cases}
\tag{4.57}
$$

– the transition rate can be a power of the considered variable. Hatzis proposes adopting a Weibull distribution, which leads to a transition law of the type $\Gamma(x) = \lambda c(\lambda x)^{c-1}$.

By considering three phases of the cell cycle, the authors end up with the following system which actually combines three population balances (one for each cycle's phase). In practice, the modeled situation is as follows: in the cycle's first phase, the cells do not vary in mass, only their age varies; in the cycle's second phase, the cells gain in mass and the transition to phase 3 is entirely controlled by a criterion based on mass. Phase 3 is also a phase without mass variation. Phase 1 receives the daughter cells resulting from the division which takes place during the last phase of the cycle. Such a proposal makes sense with respect to the curve of the cell length's evolution shown in Figure 4.8,

$$
\frac{\partial N_1(x,t)}{\partial t} + \frac{\partial}{\partial x_i}\left[u_{1_i}.N_1(x,t)\right] = 2\int_x^\infty \Gamma_3(x,t)p(x,x')N_3(x',t)dx' - \Gamma_1(x,t)N_1(x,t)
$$

$$
\frac{\partial N_2(x,t)}{\partial t} + \frac{\partial}{\partial x_i}\left[u_{2_i}.N_2(x,t)\right] = \Gamma_1(x,t)N_1(x,t) - \Gamma_2(x,t)N_2(x,t)
$$

$$
\frac{\partial N_3(x,t)}{\partial t} + \frac{\partial}{\partial x_i}\left[u_{3_i}.N_3(x,t)\right] = \Gamma_2(x,t)N_2(x,t) - \Gamma_3(x,t)N_3(x,t)
$$

$$
\tag{4.58}
$$

where:

– x is the vector of internal variables such as mass and age;

– u_{ki} is the velocity in space of internal variable i for phase k of the cycle.

The terms on the right-hand side correspond to source terms in the fundamental population balance equation. From a certain point of view, this modeling can be regarded as the juxtaposition of three population balances on different variables (age or mass). The mass distribution does not evolve in phase 1 and the transition frequency Γ_1 to phase 2 depends only on age. Hence, vector x in the equation of $N1$ is reduced to the age. It will be noted that the displacement rate u_1 in the age space is independent of age because, by definition, all cells age by one second per second. The second balance concerning N_2 describes the evolution of a mass distribution without reference to age. In the same way, the balance on N_3 relates again only to the cells',

$$\frac{\partial N_1(a,m,t)}{\partial t} + \frac{\partial}{\partial a}[1.N_1(a,m,t)] = -\Gamma_1(a)N_1(a,m,t)$$

$$\frac{\partial N_2(m,t)}{\partial t} + \frac{\partial}{\partial m}[u_2(m).N_2(m,t)] = \int_0^\infty \Gamma_1(a)N_1(a,m,t)da - \Gamma_2(m)N_2(m,t)$$

$$\frac{\partial N_3(a,m,t)}{\partial t} + \frac{\partial}{\partial a}[1.N_3(a,m,t)] = -\Gamma_3(a)N_3(a,m,t)$$

[4.59]

This set of equations is supplemented by the following limit conditions:

– cells of age 0 and mass m in the cycle's first phase come from the division of the cells in the cycle's third phase and from mass $m' > m$

$$N_1(0,m,t) = 2\int_0^\infty \int_m^\infty \Gamma_3(a)N_3(a,m,t)p(m,m')dm'da$$

[4.60]

– the cells that enter phase 3 with a mass m are those that leave phase 2 with this same mass. It should be noted that the age is 0 at the beginning of phase 3. This is therefore an age per cycle phase and not an age in the complete cell cycle,

$$N_3(0,m,t) = \Gamma_2(m)N_2(m,t)$$

[4.61]

4.3.2.2. Resolution

The method used is the discrete event Monte Carlo method. It typically involves randomly drawing the probabilities that a given event occurs within a given time period and applying it to a randomly selected cell in the population. This is a statistical method because it is necessary to take into account the previous drawings and modify the probabilities of the different events according to the drawings. In this approach, the population is effectively represented by a set of individuals. The average properties at the population scale will be obtained by averaging all the realizations on all individuals. This vision forms the basis of an entire class of population models called "cell ensemble model".

The main disadvantage arises from the fact that cell multiplication logically leads to an exponential growth in the number of individuals, the number of draws to be made and thus the duration of the simulation. Improvements have been made by reasoning on a constant number of cells (constant number Monte Carlo) [MAN 06]. After each division, the procedure is completed by the elimination of a random cell.

4.3.2.3. Results

Having chosen a constant growth rate $u_2(m) = k$, and a distribution law of the mass during the division according to a beta distribution (with parameter q = 40), the following results are obtained (Figure 4.9):

$$p(m, m') = \frac{1}{B(q,q)} \cdot \frac{1}{m\prime} \left(\frac{m}{m'}\right)^{q-1} \left(1 - \frac{m}{m'}\right)^{q-1} \qquad [4.62]$$

After erasing the initial conditions, the numbers of individuals in each phase evolve exponentially over time and the fractions in each phase become constant. N_1 populations less than 5,000 and N_3 greater than the minimum critical mass (m_c = 5,000, arbitrary units) are clearly seen on graph a. On graph b, population N_2 is shown to represent, in the studied case, 90% of the total population in number. The mass asymmetric Gaussian-type distribution corresponds to experimental observations.

NOTE.– The model corresponds to unlimited growth in a constant environment. The population growth in number is exponential. After eliminating the initial conditions, the mass distribution becomes independent of time. There is therefore also an exponential growth in mass. This result is obtained even though the growth rate of all individuals is identical, independent of mass. Adding a temporal dimension to the model has thus made it possible to reconcile the observation on the elongation of the Nobs and Maerkl cells (Figure 4.7, linear variation in the length over time) and the macroscopic observation of an exponential growth in mass.

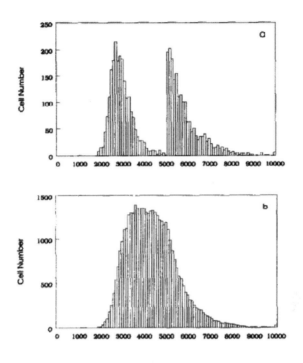

Figure 4.9. *Number distributions in phases 1 and 3 (graph a)
and in phase 2 (graph b) at long times: in exponential
growth regime [HAT 95]*

4.3.2.4. *Experimental determination of nuclei* Γ

Recently, a team of INRIA researchers proposed an experimental validation of this type of modeling based on the in-line measurement of cell fluorescence [BIL 13]. The technique called FUCCI (Fluorescence Ubiquitination-based Cell Cycle Indicator) is based on the development of two fluorescent probes fused with two proteins, whose concentration varies during the cell cycle. One emits in red during the G_1 phase and the other in green during the other three phases. The technique has been applied to mouse embryo fibroplasts. The population balance used is very similar to that presented above; it is solved using a finite difference-type method.

These analyses make it possible to define the probability that a given cell passes a time τ in a phase of the cycle and to deduce the transition frequencies Γ present in the model.

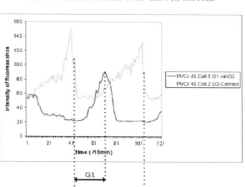

Figure 4.10. *Graphical representation of the method used to determine the duration of the G1 phase and the complete cell cycle. The total duration of the S/G2/M phases is deduced by subtraction. For a color version of this figure, see www.iste.co.uk/morchain/bioreactor.zip*

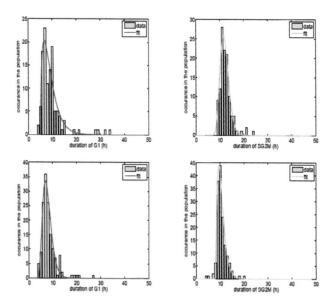

Figure 4.11. *Gamma laws for the G1 (left) and S/G2/M (right) phases for two experimental conditions: 10% FBS (top) and 15% FBS (bottom). For a color version of this figure, see www.iste.co.uk/morchain/bioreactor.zip*

Figure 4.11 shows that the frequencies depend on the medium's composition in terms of the concentration of growth factors (fetal bovine serum or FBS).

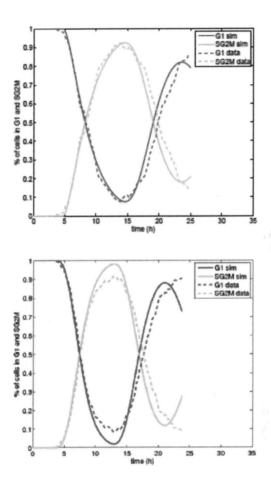

Figure 4.12. *Temporal evolution of the percentage of cells in phases G1 and S/G2/M. Experimental data are shown in dotted lines, 10% FBS (top) and 15% FBS (bottom), and simulations in continuous line. For a color version of this figure, see www.iste.co.uk/morchain/bioreactor.zip*

As in the works of Hatzis *et al.* [HAT 95], it is observed that the initially synchronous population (100% of individuals in phase G1) desynchronizes at a different rate depending on the medium's composition, leading to a constant proportion of individuals in the two phases at the end of the experiment.

4.3.3. Differentiation by intracellular composition

It is quite clear that if we want to describe the cellular behavior with more precision, it is necessary to increase the number of internal variables. This is especially the case as soon as we seek to describe the evolution of the intracellular concentrations of the compounds involved in a metabolic pathway. The choice of the method of resolution then becomes critical.

The class-based method [KUM 96a, KUM 96b, KUM 97] reaches its limits because each direction of the internal space has to be discretized with a large number of points (typically about one hundred). Thus, the number of equations of the complete model increases according to a power law as a function of the number of internal variables p. We obtain a number of equations in N^p. The resolution becomes prohibitive in terms of computing time. Henson has shown that this conclusion extends to other techniques such as finite differences, finite elements or orthogonal collocation for the same reasons [HEN 05]. This author therefore proposes a so-called "cell ensemble modeling" approach. The idea consists of using an identical cellular model, based on a reduced number of internal variables (6 in the cited work but can in fact be much larger), applied to a large number of cells (> 1,000). From a heterogeneous population (in composition), this type of model converges to a state where all the cells evolve synchronously. This means that they eventually converged to the same state. By adding a random perturbation of the reaction rates or the composition, sustained oscillations can be obtained [HEN 03a]. It is then necessary to use a structured model in which the values of the reaction rates are drawn randomly using a Gaussian-type distribution law for the intracellular reaction rates. It can be noted that these perturbations play a similar role to the functions describing cell division, Γ, used in a classical approach using a population balance equation. For a fairly complete overview of the techniques used today to solve this type of problem, we can read (and re-read) the work of Stamatakis [STA 10], which mentions, among others, the 3D and 6D population balances.

4.3.4. Coupling with the bioreactor

4.3.4.1. Ideal bioreactor model

The number of studies in this area is quite small [HEI 15, HEN 03b, HEN 03a, HEN 05, KOL 12, MOR 13, MOR 16, PIG 15]. To make a coupling between the population balance model and the environment, that is, the liquid-phase concentrations. It is necessary to make the functions (typically u and Γ in the preceding models) depend on the composition of the liquid phase. Moreover, the consumption rate in the liquid phase is now calculated as an integral on the set of rates realized by the individuals (see, for example, equation [4.33] at the beginning of this chapter). The coupling then becomes more effective: the temporal variations imposed in the environment can be reflected in the population. Everything then depends on the biological phenomena taken into account and on the relationship between the associated characteristic time and that of the external fluctuations undergone in relation to the hydrodynamics and the reactor's mixing state. However, a distinction should be made between:

– studies of a single reactor for which a large number of internal variables can be authorized and a wide range of resolution methods used;

– studies that consider the combination of several ideal reactors [HEI 15, PIG 15]. The need to transport the density function in physical space greatly reduces the number of internal variables and the spectrum of usable numerical methods. The two cited studies use the class-based method and a one-dimensional population balance coupled with a metabolic model.

We propose a population balance model for the distribution of specific growth rates of individuals coupled with an unstructured kinetic model. The advantages of choosing the growth rate μ^b as an internal variable are many:

– this variable combines the two scales, mass and time. It reflects the potential growth rate of individuals which is fully expressed unless substrate is missing;

– the information on the potential growth rate of individuals makes it possible to calculate the substrate demand;

– the information can be used to reduce the level of sub-determination of a metabolic model whose resolution then becomes algebraic;

– comparing with the achievable growth rate, given the concentrations in liquid phase, it allows determining which individuals fully express their potential and those for whom the medium is limiting.

The model's equations are given by expressions [4.61]. The density function corresponds to the mass distribution of the cell able to grow at μ^b (per unit volume). Therefore, it is directly expressed in terms of cell concentration,

$$\frac{\partial X(\mu^b)}{\partial t} + \frac{\partial}{\partial \mu}\left(u_\mu X(\mu^b, t)\right) = \mu^r X(\mu^b, t)$$

$$u_\mu = \left(\frac{1}{\tau_\mu} + \mu^b\right)\left(\frac{\mu^b - \tilde{\mu}^*}{\tilde{\mu}_{max}}\right)$$

$$\tilde{\mu}^* = \tilde{\mu}_{max}\frac{S}{K_S + S} \qquad\qquad [4.63]$$

$$\frac{\partial S}{\partial t} = \frac{Q}{V}\left(S_f - S\right) - \int_0^\infty \frac{1}{Y_{XS}(\mu)}\mu^r X(\mu, t)d\mu$$

$$\mu^r = \min(\mu^b, \tilde{\mu}^*)$$

The second member of the first equation relies on the assumption that daughter cells retain the same specific growth rates as their mother. The adaptation rate of the growth rate u_μ corresponds to that already presented in Chapter 3 (equation [3.50]). It indicates that the individuals' growth rate relaxes towards the equilibrium value $\tilde{\mu}^*(S)$, which would be that of the population acclimated to the concentration S. The last equation is a material balance on the substrate in an open reactor. Here, the conversion yield is assumed to vary with the growth rate of individuals. The last equation indicates that the individuals' real growth rate is possibly limited by the substrate's availability in the liquid phase. Individuals only realize their potential growth rate if the medium is sufficiently rich. It is necessary to consider the possibility of growth being limited by the cells' capacity $\mu^r = \mu^b$ if $\mu^b < \tilde{\mu}^*$, or by the external composition $\mu^r = \tilde{\mu}^*$ if $\tilde{\mu}^* < \mu^b$.

The results presented in Figures 4.13 and 4.14 relate to the same experiment as that presented in Chapter 3 concerning a shift in the dilution rate in chemostat [KÄT 86].

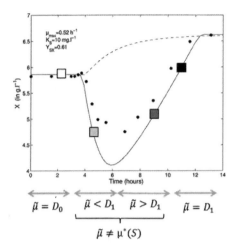

Figure 4.13. *Evolution of the biomass concentration in response to a sudden increase in the dilution rate from $D_0 = 0.1 \; h^{-1}$ to $D_1 = 0.42 \; h^{-1}$*

The partial leaching of the culture and the resumption of growth are well represented with the population balance model. In the transitional phase, the population's growth rate is not equal to the growth rate at equilibrium. If growth at equilibrium is assumed (growth rate computed from the substrate concentration), the evolution of the biomass concentration follows the dotted line.

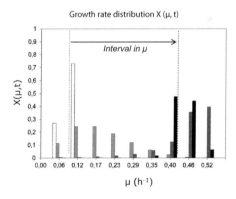

Figure 4.14. *Evolution of the distribution of growth rates during the response to a sudden increase in the dilution rate from $D_0 = 0.1 \; h^{-1}$ to $D_1 = 0.42 \; h^{-1}$*

An equilibrium-based growth model predicts an instantaneous jump of μ in response to the sharp increase in substrate concentration due to the shift of the dilution rate.

The gray level of the histograms in Figure 4.14 corresponds to the instants observed on the biomass concentration curve in Figure 4.13. The population balance model shows that the distribution of growth rates shifts gradually towards the right. This motion in the space of the μ is related to the fact that the substrate concentration has increased, so for $\tilde{\mu}^*$ also (it is then close to μ_{max}).

During the leaching phase, the distribution remains generally below D_1. As the medium is rich in substrate, the distribution continues to shift towards the right and the resulting growth rate becomes greater than the dilution rate D_1. When the substrate concentration becomes smaller, $\tilde{\mu}^*$ decreases and the distribution then slides to the left and finally stabilizes around the value D_1.

4.3.4.2. Real reactor model

One of the simplest cases to consider approaching a real reactor is the case of two perfectly mixed reactors between which the cells circulate. We present here a configuration where the substrate is fed into a reactor while oxygen is transferred to the other reactor (see Figure 4.15).

The resolved equations are those of model [4.63] with an additional Monod-type term on oxygen in the growth law at equilibrium. Of course, it is necessary to add the transport equations between the two compartments. The ADENON bioreactor dynamic simulation tool was used for class-based resolution. This simulator was developed by Jérôme Morchain and Maxime Pigou, within the Transfer-Interface-Mixture team, INSA, Toulouse.

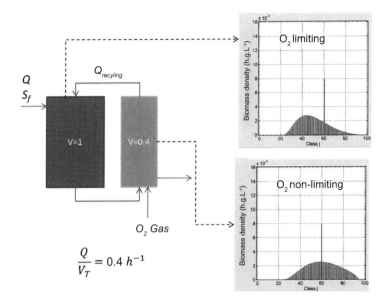

Figure 4.15. *Distribution of growth rates X(μ) in the two compartments of a multi-stage bioreactor with internal recycling. For a color version of this figure, see www.iste.co.uk/morchain/bioreactor.zip*

The growth rate imposed on the population as a whole through the overall dilution rate is 0.4 h^{-1}, which corresponds to 60% of the population's maximum growth rate. This value is indicated by a vertical red line on the graphs showing population growth rate distributions. It is observed that the population growth rate distributions are different in the two compartments. They are wide spread due to the distribution of the residence times in each of the reactors. The spread is also partly due to the fact that the adaptation characteristic time is of the same order of magnitude as the residence time in each compartment (controlled via the recycling rate). Thus, individuals[6] have a significant amount of time to adapt to local concentration conditions but not enough time to reach the state of local equilibrium with the environment specific to each compartment. In the low-oxygen zone, the local growth rate is lower than 0.4 h^{-1} and the population shows a growth

---

6 On average, when passing in a compartment.

rate that is also lower than the overall dilution rate. In the oxygen-rich zone, the local growth rate at equilibrium is close to the maximum growth rate. The population shows a growth rate well above 0.4 h^{-1}.

The recycling rate has a decisive influence on the result:

– at an infinite recycling rate, we tend towards a homogeneous reactor-type behavior. The distribution is identical in the two compartments; it is very close to the value of the overall dilution rate;

– at a recycling rate close to zero, the concentration distributions in the two environments are very different because the average residence times in each zone are sufficient for populations to adapt to local conditions.

Let us now consider the case where the reactor's internal hydrodynamics and the local concentration distributions are provided by a CFD calculation. The number of elementary volumes then reaches the order of a million. Consequently, the association of the hydrodynamic model with a multivariable population balance model solved by the methods mentioned previously (class-type, finite differences, spectral, Monte Carlo) or an ensemble model becomes impracticable. With a "cell-ensemble" approach, each elementary volume should have a set of cells large enough to be representative of the population at that location (> 1,000). Here, we come back to the limitation that is encountered if we adopt a Lagrangian view for the population's description. With a PBE approach, the cost of calculating integral terms becomes prohibitive.

To our knowledge, there is no published study combining a population balance model and CFD for the simulation of mixing/transfer/reaction interactions in biological reactors. However, such coupling has been carried out in the case of a pure plug flow reactor solved by CFD and validated by comparison with the solution obtained with the same batch model [MOR 13]. In a second study, the rate field and substrate concentration in a 70 m^3 fed-batch bioreactor was simulated by CFD. The biomass concentration (constant and unresolved) was set so that the consumption characteristic time was less than the mixing time. The result is a spatially heterogeneous concentration field (Figure 4.16).

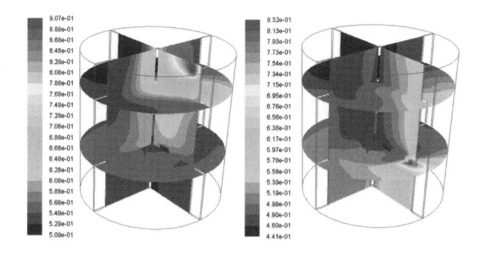

Figure 4.16. *Visualizations of the substrate concentration field (in g.l⁻¹) in a 70 m³ bioreactor. Left: feed at the surface; right: feed near the mixer [MOR 14]. For a color version of this figure, see www.iste.co.uk/morchain/bioreactor.zip*

A population balance identical to the one presented in equation [4.63] was then transported on this (stationary) concentration field. As the concentration field is stationary, we also tend towards a stationary growth rate distribution $X(\mu, x, y, z)$. The results show that the population's average growth rate at the reactor scale (average in the μ-space and in the physical space) is identical to the growth rate based on the average substrate volume concentration. Moreover, distribution $X(\mu)$ is the same at all points of the reactor. This result is consistent with the fact that the population's adaptation time is large compared with the mixing time in this reactor. However, the difference between the population's average local growth rate $\tilde{\mu}(x, y, z)$ and the local growth rate at equilibrium $\tilde{\mu}^*(x, y, z)$ varies significantly in the reactor. In other words, although being always in the same state, the population sometimes perceives areas where the amount of substrate present exceeds its needs and others where it is insufficient in relation to its needs. On the basis of this information, we have estimated the difference between the assimilation rate, based on the local substrate concentration $S(x, y, z)$, and the substrate utilization rate based on the distribution $X(\mu)$.

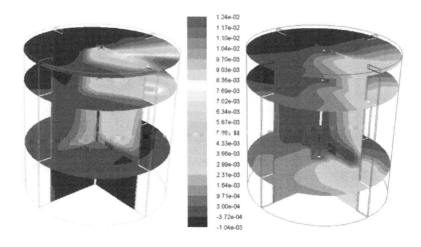

Figure 4.17. *Difference between assimilation and local substrate utilization rates.*
Effect of injection site. For a color version of this figure, see
www.iste.co.uk/morchain/bioreactor.zip

The results presented in Figure 4.17 show that, from the viewpoint of the
population of cells traveling in the bioreactor, the feeding at the surface
induces large differences between assimilation and utilization rates
(left figure). In a large area around the injection site, the cells over-
assimilate the substrate in relation to their need. Far from the injection site,
assimilation does not cover the cells' needs (negative difference). The
difference may seem small, but the fraction of the reactor's volume involved
in this situation is very large. Thus, the cells are globally out of equilibrium
and their feed is chaotic: sometimes in excess and sometimes insufficient.
This repeated exposure to imbalances between substrate assimilation and
utilization rates is certainly the cause of the metabolic drifts observed in
large reactors. In the case of a feed point close to the mixer, the substrate
dispersion is much better (right figure). From the cells' point of view,
concentration fluctuations are much smaller and it is legitimate to argue that
the risk of a metabolic drift is reduced [MOR 14].

The ADENON simulation tool was used to simulate the extrapolation of a
process between the 20 l and 20 m^3 scale. A full coupling of the equations of
transport, transfer, biological reaction and dynamic adaptation of the
individuals' growth rate was realized. A metabolic model for *E. coli*
integrating dissimilation was coupled with the population balance model

[PIG 15]. The reactor is represented by 70 interconnected compartments following the methodology established by Vrábel *et al.* [VRÁ 99, VRÁ 00, VRÁ 01]. The obtained results confirm that the area of greatest imbalance is the site of an acetate-producing overflow-type metabolism, whereas low-glucose areas from the point of view of the population's needs induce a re-consumption of acetate (Figure 4.18). This "futile" dynamic explains the loss of biomass yield during extrapolation [PIG 15].

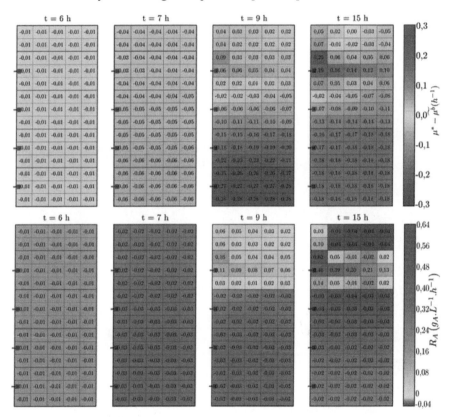

Figure 4.18. *Mapping of the population gap at equilibrium and the average acetate production/consumption rate by the cell population. For a color version of this figure, see www.iste.co.uk/morchain/bioreactor.zip*

Bibliography

[ABU 89] ABULESZ E.-M., LYBERATOS G., "Periodic operation of a continuous culture of Baker's yeast", *Biotechnology and Bioengineering*, vol. 34, pp. 741–749, 1989. doi: 10.1002/bit.260340603.

[ADA 09] ADAMBERG K., LAHTVEE P.-J., VALGEPEA K. *et al.*, "Quasi steady state growth of Lactococcus lactis in glucose-limited acceleration stat (A-stat) cultures", *Antonie Van Leeuwenhoek*, vol. 95, pp. 219–226, 2009. doi: 10.1007/s10482-009-9305-z.

[AL 07] AL-HOMOUD A., HONDZO M., LAPARA T., "Fluid dynamics impact on bacterial physiology: biochemical oxygen demand", *Journal of Environmental Engineering*, vol. 133, pp. 226–236, 2007.

[AMA 00] AMANULLAH A., JÜSTEN P., DAVIES A. *et al.*, "Agitation induced mycelial fragmentation of Aspergillus oryzae and Penicillium chrysogenum", *Biochemical Engineering Journal*, vol. 5, p. 109, 2000.

[AMA 01] AMANULLAH A., MCFARLANE C.M., EMERY A.N. *et al.*, "Scale-down model to simulate spatial pH variations in large-scale bioreactors", *Biotechnology and Bioengineering*, vol. 73, pp. 390–399, 2001.

[AUG 03] AUGIER F., MORCHAIN J., GUIRAUD P. *et al.*, "Volume fraction gradient-induced flow patterns in a two-liquid phase mixing layer", *Chemical Engineering Science*, vol. 58, pp. 3985–3993, 2003. doi: 10.1016/S0009-2509(03)00267-7.

[BAI 86] BAILEY J.E., OLLIS D.F., *Biochemical Engineering Fundamentals*, McGraw Hill, Singapore, 1986.

[BAJ 82] BAJPAI R.K., REUSS M., "Coupling of mixing and microbial kinetics for evaluating the performance of bioreactors", *The Canadian Journal of Chemical Engineering*, vol. 60, pp. 384–392, 1982. doi: 10.1002/cjce.5450600308.

[BAL 90] BALDYGA J., BOURNE J.R., "Comparison of the engulfment and the interaction-by-exchange-with-the-mean micromixing models", *Chemical Engineering Science*, vol. 45, pp. 25–31, 1990.

[BAL 97] BALDYGA J., BOURNE J.R., HEARN S.J., "Interaction between chemical reactions and mixing on various scales", *Chemical Engineering Science*, vol. 52, pp. 457–466, 1997.

[BAL 03] BALDYGA J., BOURNE J.R., *Turbulent Mixing and Chemical Reactions*, John Wiley and Sons, Chichester, 2003.

[BAR 79] BARFORD J.P., HALL R.J., "An examination of the Crabtree effect in *Saccharomyces cerevisiae*: the role of respiratory adaptation", *Microbiology*, vol. 114, pp. 267–275, 1979.

[BEN 06] BEN CHAABANE F., ALDIGUIER A.S., ALFENORE S. *et al.*, "Very high ethanol productivity in an innovative continuous two-stage bioreactor with cell recycle", *Bioprocess and Biosystems Engineering*, vol. 29, pp. 49–57, 2006. doi: 10.1007/s00449-006-0056-1.

[BER 10] BEROVIČ M., ENFORS S.O., Comprehensive bioprocess engineering, Thesis, University of Ljubljana, 2010.

[BEZ 03] BEZZO F., MACCHIETTO S., PANTELIDES C.C., "General hybrid multizonal/CFD approach for bioreactor modeling", *AIChE Journal*, vol. 49, pp. 2133–2148, 2003.

[BEZ 05] BEZZO F., MACCHIETTO S., PANTELIDES C.C., "Computational issues in hybrid multizonal/computational fluid dynamics models", *AIChE Journal*, vol. 51, pp. 1169–1177, 2005.

[BIL 13] BILLY F., CLAIRAMBAULT J., DELAUNAY F. *et al.*, "Age-structured cell population model to study the influence of growth factors on cell cycle dynamics", *Mathematical Biosciences and Engineering*, vol. 10, pp. 1–17, 2013. doi: 10.3934/mbe.2013.10.1.

[BOY 03] BOYE E., NORDSTROM K., "Coupling the cell cycle to cell growth", *EMBO Reports*, vol. 4, pp. 757–760, 2003.

[BYL 98] BYLUND F., COLLET E., ENFORS S.-O. *et al.*, "Substrate gradient formation in the large-scale bioreactor lowers cell yield and increases by-product formation", *Bioprocess and Biosystems Engineering*, vol. 18, pp. 171–180, 1998.

[BYL 99] BYLUND F., GUILLARD F., ENFORS S.-O. *et al.*, "Scale down of recombinant protein production: a comparative study of scaling performance", *Bioprocess Engineering*, vol. 20, pp. 377–389, 1999. doi: 10.1007/s004490050606.

[CAR 04] CARLSON R., SRIENC F., "Fundamental *Escherichia coli* biochemical pathways for biomass and energy production: creation of overall flux states", *Biotechnology and Bioengineering*, vol. 86, pp. 149–162, 2004.

[CHA 02] CHASSAGNOLE C., NOISOMMIT-RIZZI N., SCHMID J.W. *et al.*, "Dynamic modeling of the central carbon metabolism of *Escherichia coli*", *Biotechnology and Bioengineering*, vol. 79, pp. 53–73, 2002.

[CHI 76] CHI C.T., HOWELL J.A., "Transient behavior of a continuous stirred tank biological reactor utilizing phenol as an inhibitory substrate", *Biotechnology and Bioengineering*, vol. 18, pp. 63–80, 1976. doi: 10.1002/bit.260180106.

[CHU 96] CHUNG J.D., STEPHANOPOULOS G., "On physiological multiplicity and population heterogeneity of biological systems", *Chemical Engineering Science*, vol. 51, pp. 1509–1521, 1996.

[COC 97] COCKX A., LINÉ A., ROUSTAN M. *et al.*, "Numerical simulation and physical modeling of the hydrodynamics in an air-lift internal loop reactor", *Chemical Engineering Science*, vol. 52, p. 3787, 1997.

[COO 88] COOKE M., MIDDLETON J.C., BUSH J.R., "Mixing and mass transfer in filamentous fermentations", *Proceedings of the 2nd International Conference Bioreactor Fluid Dynamics*, Cranfield, UK, pp. 37–64, 1988.

[DAN 06] DANI A., COCKX A., GUIRAUD P., "Direct numerical simulation of mass transfer from spherical bubbles: the effect of interface contamination at low Reynolds numbers", *International Journal of Chemical Reactor Engineering*, vol. 4, pp. 1–21, 2006.

[DAN 07] DANI A., GUIRAUD P., COCKX A., "Local measurement of oxygen transfer around a single bubble by planar laser-induced fluorescence", *Chemical Engineering Science*, vol. 62, p. 7245, 2007.

[DEL 05] DELVIGNE F., DESTAIN J., THONART P., "Structured mixing model for stirred bioreactors: an extension to the stochastic approach", *Chemical Engineering Journal*, vol. 113, p. 1, 2005.

[DEL 08a] DELAFOSSE A., Analyse et étude numérique des effets de mélange dans un bioréacteur, Thesis, Institut National des Sciences Appliquées, Toulouse, France, 2008.

[DEL 08b] DELVIGNE F., DESTAIN J., THONART P., "Elaboration of a biased random walk model with a high spatial resolution for the simulation of the microorganisms exposure to gradient stress in scale-down reactors", *Biochemical Engineering Journal*, vol. 39, p. 105, 2008.

[DEL 09] DELAFOSSE A., MORCHAIN J., GUIRAUD P., "Mixing study in a bioreactor by LES and particle tracking", *8th World Congress of Chemical Engineering*, Montreal, Quebec, 2009.

[DEL 10] DELVIGNE F., INGELS S., THONART P., "Evaluation of a set of *E. coli* reporter strains as physiological tracer for estimating bioreactor hydrodynamic efficiency", *Process Biochemistry*, vol. 45, pp. 1769–1778, 2010.

[DEL 11] DELVIGNE F., BROGNAUX A., GORRET N. *et al.*, "Characterization of the response of GFP microbial biosensors sensitive to substrate limitation in scale-down bioreactors", *Biochemical Engineering Journal*, vol. 55, pp. 131–139, 2011.

[DEL 15] DELVIGNE F., BAERT J., GOFFLOT S. *et al.*, "Dynamic single-cell analysis of *Saccharomyces cerevisiae* under process perturbation: comparison of different methods for monitoring the intensity of population heterogeneity", *Journal of Chemical Technology and Biotechnology*, vol. 90, pp. 314–323, 2015. doi: 10.1002/jctb.4430.

[DIA 10] DIAZ M., HERRERO M., GARCIA L.A. *et al.*, "Application of flow cytometry to industrial microbial bioprocesses", *Biochemical Engineering Journal*, vol. 48, pp. 385–407, 2010.

[DUN 90] DUNLOP E.H., YE S.J., "Micromixing in fermentors: metabolic changes in *Saccharomyces cerevisiae* and their relationship to fluid turbulence", *Biotechnology and Bioengineering*, vol. 36, pp. 854–864, 1990.

[ENF 01] ENFORS S.O., JAHIC M., ROZKOV A. *et al.*, "Physiological responses to mixing in large scale bioreactors", *Journal of Biotechnology,* vol. 85, pp. 175–185, 2001.

[ESE 83] ESENER A.A., ROELS J.A., KOSSEN N.W.F., "Theory and applications of unstructured growth models: kinetic and energetic aspects", *Biotechnology and Bioengineering*, vol. 25, pp. 2803–2841, 1983. doi: 10.1002/bit.260251202.

[FAN 70] FAN L.T., ERICKSON L.E., SHAH P.S. *et al.*, "Effect of mixing on the washout and steady-state performance of continuous cultures", *Biotechnology and Bioengineering*, vol. 12, pp. 1019–1068, 1970. doi: 10.1002/bit.260120611.

[FAN 71] FAN L.T., TSAI B.I., ERICKSON L.E., "Simultaneous effect of macromixing and micromixing on growth processes", *AIChE Journal*, vol. 17, pp. 689–696, 1971. doi: 10.1002/aic.690170336.

[FER 96] FERENCI T., "Adaptation to life at micromolar nutrient levels: the regulation of *Escherichia coli* glucose transport by endoinduction and cAMP", *FEMS Microbiology Reviews*, vol. 18, pp. 301–317, 1996.

[FER 99a] FERENCI T., "Growth of bacterial cultures' 50 years on: towards an uncertainty principle instead of constants in bacterial growth kinetics", *Research in Microbiology*, vol. 150, pp. 431–438, 1999.

[FER 99b] FERENCI T., "Regulation by nutrient limitation", *Current Opinion in Microbiology*, vol. 2, pp. 208–213, 1999.

[FER 01] FERENCI T., "Hungry bacteria – definition and properties of a nutritional state", *Environmental Microbiology*, vol. 3, pp. 605–611, 2001. doi: 10.1046/j.1462-2920.2001.00238.x.

[FER 07] FERENCI T., ROBERT K.P., "Bacterial physiology, regulation and mutational adaptation in a chemostat environment", *Advances in Microbial Physiology*, Academic Press, pp. 169–229, 2007.

[FOU 96] FOURNIER M.-C., FALK L., VILLERMAUX J., "New parallel competing reaction system for assessing micromixing efficiency – experimental approach", *Chemical Engineering Science*, vol. 51, pp. 5053–5064, 1996. doi: 10.1016/0009-2509(96)00270-9.

[FOX 03] FOX R.O., *Computational Models for Turbulent Reacting Flows*, Cambridge University Press, Cambridge, 2003.

[FRE 63] FREDRICKSON A.G., TSUCHIYA H.M., "Continuous propagation of microorganisms", *AIChE Journal*, vol. 9, pp. 459–468, 1963.

[FUE 13] FUENTES L.G., LARA A.R., MARTINEZ L.M. *et al.*, "Modification of glucose import capacity in *Escherichia coli*: physiologic consequences and utility for improving DNA vaccine production", *Microbial Cell Factories*, vol. 12, 2013.

[GAB 12] GABELLE J.-C., Analyse locale et globale de l'hydrodynamique et du transfert de matière dans des fluides à rhéologie complexe caractéristiques des milieux de fermentation, PhD Thesis, Institut National des Sciences Appliquées, Toulouse, France, 2012.

[GAB 13] GABELLE J.-C., MORCHAIN J., ANNE-ARCHARD D. *et al.*, "Experimental determination of the shear rate in a stirred tank with a non-Newtonian fluid: Carbopol", *AIChE Journal*, vol. 59, pp. 2251–2266, 2013. doi: 10.1002/aic.13973.

[GAR 09] GARCIA J., CHA H., RAO G. *et al.*, "Microbial nar-GFP cell sensors reveal oxygen limitations in highly agitated and aerated laboratory-scale fermentors", *Microbial Cell Factories*, vol. 8, p. 6, 2009. doi: 10.1186/1475-2859-8-6.

[GUI 04] GUILLOU V., PLOURDE-OWOBI L., PARROU J.L. *et al.*, "Role of reserve carbohydrates in the growth dynamics of *Saccharomyces cerevisiae*", *FEMS Yeast Research*, vol. 4, pp. 773–787, 2004. doi: 10.1016/j.femsyr.2004.05.005.

[HAN 66] HANSFORD G.S., HUMPHREY A.E., "The effect of equipment scale and degree of mixing on continuous fermentation yield at low dilution rates", *Biotechnology and Bioengineering*, vol. 8, pp. 85–96, 1966. doi: 10.1002/bit.260080108.

[HAR 16] HARINGA C., TANG W., DESHMUKH A.T. *et al.*, "Euler-Lagrange computational fluid dynamics for (bio)reactor scale down: an analysis of organism lifelines", *Engineering in Life Sciences*, vol. 16, pp. 652–663, 2016. doi: 10.1002/elsc.201600061.

[HAT 95] HATZIS C., SRIENC F., FREDRICKSON A.G., "Multistaged corpuscular models of microbial growth: Monte Carlo simulations", *Biosystems*, vol. 36, pp. 19–35, 1995.

[HEI 05] HEIJNEN J.J., "Approximative kinetic formats used in metabolic network modeling", *Biotechnology and Bioengineering*, vol. 91, pp. 534–545, 2005.

[HEI 15] HEINS A.-L., LENCASTRE FERNANDES R., GERNAEY K.V. *et al.*, "Experimental and *in silico* investigation of population heterogeneity in continuous *Sachharomyces cerevisiae* scale-down fermentation in a two-compartment setup", *Journal of Chemical Technology and Biotechnology*, vol. 90, pp. 324–340, 2015. doi: 10.1002/jctb.4532.

[HEN 03a] HENSON M.A., "Dynamic modeling and control of yeast cell populations in continuous biochemical reactors", *Computers & Chemical Engineering*, vol. 27, pp. 1185–1199, 2003.

[HEN 03b] HENSON M.A., "Dynamic modeling of microbial cell populations", *Current Opinion in Biotechnology*, vol. 14, pp. 460–467, 2003.

[HEN 05] HENSON M.A., "Cell ensemble modeling of metabolic oscillations in continuous yeast cultures", *Computers & Chemical Engineering*, vol. 29, pp. 645–661, 2005.

[HEW 99] HEWITT C.J., NEBE-VON CARON G., NIENOW A.W. *et al.*, "The use of multi-parameter flow cytometry to compare the physiological response of *Escherichia coli* W3110 to glucose limitation during batch, fed-batch and continuous culture cultivations", *Journal of Biotechnology*, vol. 75, p. 251, 1999.

[HJE 95] HJERTAGER B., MORUD K., "Computational fluid dynamics simulation of bioreactors", *Modeling, Identification and Control*, vol. 16, pp. 177–191, 1995.

[HRI 01] HRISTOV H., MANN R., LOSSEV V. *et al.*, "A 3-D analysis of gas-liquid mixing, mass transfer and bioreaction in a stirred bio-reactor", *Food and Bioproducts Processing*, vol. 79, p. 232, 2001.

[HRI 04] HRISTOV H., MANN R., LOSSEV V. *et al.*, "A simplified CFD for three-dimensional analysis of fluid mixing, mass transfer and bioreaction in a fermenter equipped with triple novel geometry impellers", *Food and Bioproducts Processing*, vol. 82, pp. 21–34, 2004.

[HUC 09] HUCHET F., LINÉ A., MORCHAIN J., "Evaluation of local kinetic energy dissipation rate in the impeller stream of a Rushton turbine by time-resolved PIV", *Chemical Engineering Research and Design*, vol. 87, pp. 369–376, 2009.

[JON 99] JONES K.D., KOMPALA D.S., "Cybernetic model of the growth dynamics of *Saccharomyces cerevisiae* in batch and continuous cultures", *Journal of Biotechnology*, vol. 71, pp. 105–131, 1999.

[KÄT 86] KÄTTERER L., ALLEMANN H., KÄPPELI O. *et al.*, "Transient responses of continuously growing yeast cultures to dilution rate shifts: a sensitive means to analyze biology and the performance of equipment", *Biotechnology and Bioengineering*, vol. 28, pp. 146–150, 1986.

[KOC 82] KOCH A.L., HOUSTON WANG C., "How close to the theoretical diffusion limit do bacterial uptake systems function?", *Archives of Microbiology*, vol. 131, pp. 36–42, 1982.

[KOL 12] KOLEWE M.E., ROBERTS S.C., HENSON M.A., "A population balance equation model of aggregation dynamics in Taxus suspension cell cultures", *Biotechnology and Bioengineering*, vol. 109, p. 472, 2012.

[KOV 98] KOVAROVA-KOVAR K., EGLI T., "Growth kinetics of suspended microbial cells: from single-substrate-controlled growth to mixed-substrate kinetics", *Microbiology and Molecular Biology Reviews*, vol. 62, pp. 646–666, 1998.

[KRE 04] KREMLING A., FISCHER S., SAUTER T. *et al.*, "Time hierarchies in the *Escherichia coli* carbohydrate uptake and metabolism", *Biosystems*, vol. 73, pp. 57–71, 2004. doi: 10.1016/j.biosystems.2003.09.001.

[KRO 08] KROMMENHOEK E.E., VAN LEEUWEN M., WALTER H.G., "Lab-scale fermentation tests of microchip with integrated electrochemical sensors for pH, temperature, dissolved oxygen and viable biomass concentration", *Biotechnology and Bioengineering*, vol. 99, pp. 884–892, 2008.

[KRU 96] KRUIS F.E., FALK L., "Mixing and reaction in a tubular jet reactor: a comparison of experiments with a model based on a prescribed PDF", *Chemical Reaction Engineering: From Fundamentals to Commercial Plants and Products*, vol. 51, pp. 2439–2448, 1996. doi: 10.1016/0009-2509(96)00100-5.

[KUM 96a] KUMAR S., RAMKRISHNA D., "On the solution of population balance equations by discretization--I. A fixed pivot technique", *Chemical Engineering Science*, vol. 51, pp. 1311–1332, 1996.

[KUM 96b] KUMAR S., RAMKRISHNA D., "On the solution of population balance equations by discretization--II. A moving pivot technique", *Chemical Engineering Science*, vol. 51, pp. 1333–1342, 1996.

[KUM 97] KUMAR S., RAMKRISHNA D., "On the solution of population balance equations by discretization--III. Nucleation, growth and aggregation of particles", *Chemical Engineering Science*, vol. 52, pp. 4659–4679, 1997.

[LAP 04] LAPIN A., MULLER D., REUSS M., "Dynamic behavior of microbial populations in stirred bioreactors simulated with Euler-Lagrange methods: traveling along the lifelines of single cells", *Industrial & Engineering Chemistry Research*, vol. 43, pp. 4647–4656, 2004.

[LAP 06] LAPIN A., SCHMID J., REUSS M., "Modeling the dynamics of E. coli populations in the three-dimensional turbulent field of a stirred-tank bioreactor – a structured-segregated approach", *Chemical Engineering Science*, vol. 61, pp. 4783–4797, 2006.

[LAR 06] LARA A.R., GALINDO E., RAMIREZ O.T. *et al.*, "Living with heterogeneities in bioreactors: understanding the effect of environmental gradients in cells", *Molecular Biotechnology*, vol. 34, pp. 355–381, 2006.

[LAR 09] LARA A.R., TAYMAZ-NIKEREL H., MASHEGO M.R. *et al.*, "Fast dynamic response of the fermentative metabolism of *Escherichia coli* to aerobic and anaerobic glucose pulses", *Biotechnology and Bioengineering*, vol. 104, pp. 1153–1161, 2009. doi: 10.1002/bit.22503.

[LEB 74] LEBOWITZ J.L., RUBINOW S.I., "A theory for the age and generation time distribution of a microbial population", *Journal of Mathematical Biology*, vol. 1, pp. 17–36, 1974. doi: 10.1007/BF02339486.

[LEE 82] LEEGWATER M.P.M., NEIJSSEL O.M., TEMPEST D.W., "Aspects of microbial physiology in relation to process control", *Journal of Chemical Technology and Biotechnology*, vol. 32, pp. 92–99, 1982. doi: 10.1002/jctb.5030320113.

[LEN 98] LENDENMANN U., EGLI T., "Kinetic models for the growth of *Escherichia coli* with mixtures of sugars under carbon-limited conditions", *Biotechnology and Bioengineering*, vol. 59, pp. 99–107, 1998.

[LEV 72] LEVENSPIEL O., *Chemical Reaction Engineering*, John Wiley and Sons, 1972.

[LI 82] LI W.K.W., "Estimating heterotrophic bacterial productivity by inorganic radiocarbon uptake: importance of establishing time courses of uptake", *Marine Ecology Progress Series*, vol. 8, pp. 167–172, 1982.

[LI 12] LI M., XU J., ROMERO-GONZALEZ M. *et al.*, "Single cell Raman spectroscopy for cell sorting and imaging", *Current Opinion in Biotechnology*, vol. 23, pp. 56–63, 2012. doi: 10.1016/j.copbio.2011.11.019.

[LIN 00] LIN Y.H., NEUBAUER P., "Influence of controlled glucose oscillations on a fed-batch process of recombinant *Escherichia coli*", *Journal of Biotechnology*, vol. 79, pp. 27–37, 2000.

[LIN 01] LIN H.Y., MATHISZIK B., XU B. *et al.*, "Determination of the maximum specific uptake capacities for glucose and oxygen in glucose-limited fed-batch cultivations of *Escherichia coli*", *Biotechnology and Bioengineering*, vol. 73, pp. 347–357, 2001.

[LIN 12a] LINKÈS M., MARTINS AFONSO M., FEDE P. *et al.*, "Numerical study of substrate assimilation by a microorganism exposed to fluctuating concentration", *Chemical Engineering Science*, vol. 81, pp. 8–19, 2012. doi: 10.1016/j.ces.2012.07.003.

[LIN 12b] LINKÈS M., Simulation numérique et modélisation de l'assimilation de substrat par des microorganismes dans un écoulement turbulent, Thesis, Institut National Polytechnique de Toulouse, Toulouse, France, 2012.

[LIN 14] LINKÈS M., FEDE P., MORCHAIN J. *et al.*, "Numerical investigation of subgrid mixing effects on the calculation of biological reaction rates", *Chemical Engineering Science*, vol. 116, pp. 473–485, 2014. doi: 10.1016/j.ces.2014.05.005.

[LLA 08] LLANERAS F., PICÓ J., "Stoichiometric modelling of cell metabolism", *Journal of Bioscience and Bioengineering*, vol. 105, pp. 1–11, 2008.

[LOO 05] LOOSER V., HAMMES F., KELLER M. *et al.*, "Flow-cytometric detection of changes in the physiological state of E. coli expressing a heterologous membrane protein during carbon-limited fedbatch cultivation", *Biotechnology and Bioengineering*, vol. 92, pp. 69–78, 2005.

[MAN 95] MANN R., PILLAI S.K., EL-HAMOUZ A.M. *et al.*, "Computational fluid mixing for stirred vessels: progress from seeing to believing", *The Chemical Engineering Journal and the Biochemical Engineering Journal*, vol. 59, pp. 39–50, 1995.

[MAN 05] MANTZARIS N.V., "A cell population balance model describing positive feedback loop expression dynamics", *Computers & Chemical Engineering*, vol. 29, pp. 897–909, 2005.

[MAN 06] MANTZARIS N.V., "Stochastic and deterministic simulations of heterogeneous cell population dynamics", *Journal of Theoretical Biology*, vol. 241, pp. 690–706, 2006.

[MAN 07] MANTZARIS N.V., "From single-cell genetic architecture to cell population dynamics: quantitatively decomposing the effects of different population heterogeneity sources for a genetic network with positive feedback architecture", *Biophysical Journal*, vol. 92, pp. 4271–4288, 2007.

[MAR 03] MARCHISIO D., PIKTURNA J.T., FOX R.O. *et al.*, "Quadrature method of moments for population-balance equations", *AIChE Journal*, vol. 49, pp. 1266–1276, 2003.

[MAR 12] MARTZOLFF A., CAHOREAU E., COGNE G. *et al.*, "Photobioreactor design for isotopic non-stationary 13C-metabolic flux analysis (INST 13C-MFA) under photoautotrophic conditions", *Biotechnology and Bioengineering*, vol. 109, pp. 3030–3040, 2012. doi: 10.1002/bit.24575.

[MAT 65] MATELES R.I., RYU D.Y., YASUDA T., "Measurement of unsteady state growth rates of micro-organisms", *Nature*, vol. 208, pp. 263–265, 1965. doi: 10.1038/208263a0.

[MAY 93] MAYR B., HORVAT P., NAGY E. *et al.*, "Mixing-models applied to industrial batch bioreactors", *Bioprocess and Biosystems Engineering*, vol. 9, pp. 1–12, 1993. doi: 10.1007/BF00389534..

[MAY 94] MAYR B., MOSER A., NAGY E. *et al.*, "Scale-up on basis of structured mixing models: a new concept", *Biotechnology and Bioengineering*, vol. 43, pp. 195–206, 1994.

[MCG 97] MCGRAW R., "Description of aerosol dynamics by the Quadrature Method of Moments", *Aerosol Science and Technology*, vol. 27, p. 255, 1997.

[MER 95] MERCHUK J.C., ASENJO J.A., "The Monod equation and mass transfer", *Biotechnology   and   Bioengineering*,   vol.   45,   pp.   91–94,   1995. doi: 10.1002/bit.260450113.

[MOI 05] MOILANEN P., LAAKKONEN M., AITTAMAA J., "Modelling fermenters with CFD", in PUIGJANER L., ESPUÑA A. (eds), *Computer Aided Chemical Engineering*, Elsevier, Netherlands, 2005.

[MOI 07] MOILANEN P., LAAKKONEN M., VISURI O. *et al.*, "Modeling local gas–liquid mass transfer in agitated viscous shear-thinning dispersions with CFD", *Industrial and Engineering Chemistry Research*, vol. 46, pp. 7289–7299, 2007.

[MON 49] MONOD J., "The growth of bacterial cultures", *Annual Review of Micro-biology*, vol. 3, pp. 371–394, 1949. doi: 10.1146/annurev.mi.03.100149.002103.

[MOR 00] MORCHAIN J., Etude et modélisation des couplages entre cinétiques physiques et biologiques dans les réacteurs de grand volume, Thesis, Institut National des Sciences Appliquées, Toulouse, 2000.

[MOR 09] MORCHAIN J., FONADE C., "A structured model for the simulation of bioreactors under transient conditions", *American Institute of Chemical Engineers Journal*, vol. 55, pp. 2973–2984, 2009. doi: 10.1002/aic.11906.

[MOR 12] MORCHAIN J., GABELLE J.-C., COCKX A., "A combination of population balance and kinetic models for bioreactor scale-up using CFD", *XXIII North American Mixing Forum*, Playa Del Carmen, Mexico, 2012.

[MOR 13a] MORCHAIN J., GABELLE J.-C., COCKX A., "Coupling of biokinetic and population balance models to account for biological heterogeneity in bioreactors", *AIChE Journal*, vol. 59, pp. 369–379, 2013. doi: 10.1002/aic.13820.

[MOR 13b] MORCHAIN J., LINKÈS M., FEDE P. *et al.*, "A coupled direct numerical simulation-Lagrangian Particle Tracking method to investigate micromixing effects in bioreactors, The European Congress of Chemical Engineering", *9th European Congress of Chemical Engineering*, The Hague, Netherlands, 2013.

[MOR 14] MORCHAIN J., GABELLE J.-C., COCKX A., "A coupled population balance model and CFD approach for the simulation of mixing issues in lab-scale and industrial bioreactors", *AIChE Journal*, vol. 60, pp. 27–40, 2014. doi: 10.1002/aic.14238.

[MOR 16] MORCHAIN J., PIGOU M., LEBAZ N., "A population balance model for bioreactors combining interdivision time distributions and micromixing concepts", *Biochemical Engineering Journal*, 2016. doi: 10.1016/j.bej.2016.09.005.

[MÜL 97] MÜLLER S., HUTTER K.J., BLEY T. *et al.*, "Dynamics of yeast cell states during proliferation and non-proliferation periods in a brewing reactor monitored by multidimensional flow cytometry", *Bioprocess and Biosystems Engineering*, vol. 17, pp. 287–293, 1997.

[NAT 99] NATARAJAN A., SRIENC F., "Dynamics of glucose uptake by single *Escherichia coli* cells", *Metabolic Engineering*, vol. 1, pp. 320–333, 1999. doi: 10.1006/mben.1999.0125.

[NAT 00] NATARAJAN A., SRIENC F., "Glucose uptake rates of single *E. coli* cells grown in glucose-limited chemostat cultures", *Journal of Microbiological Methods*, vol. 42, pp. 87–96, 2000. doi: 10.1016/S0167-7012(00)00180-9.

[NEU 95] NEUBAUER P., HÄGGSTRÖM L., ENFORS S.O., "Influence of substrate oscillations on acetate formation and growth yield in *Escherichia coli* glucose limited fed-batch cultivations", *Biotechnology and Bioengineering*, vol. 47, pp. 139–146, 1995.

[NEU 10] NEUBAUER P., JUNNE S., "Scale-down simulators for metabolic analysis of large-scale bioprocesses", *Current Opinion in Biotechnology*, vol. 21, pp. 114–121, 2010.

[NIE 92] NIELSEN J., VILLADSEN J., "Modelling of microbial kinetics", *Chemical Engineering Science*, vol. 47, pp. 4225–4270, 1992. doi: 10.1016/0009-2509(92)85104-J.

[NIE 96] NIENOW A.W., LANGHEINRICH C., STEVENSON N.C. *et al.*, "Homogenisation and oxygen transfer rates in large agitated and sparged animal cell bioreactors: some implications for growth and production", *Cytotechnology*, vol. 22, pp. 87–94, 1996, doi: 10.1007/BF00353927

[NOB 14] NOBS J.-B., MAERKL S.J., "Long-term single cell analysis of S. pombe on a microfluidic microchemostat array", *PLoS ONE*, vol. 9, p. e93466, 2014. doi: 10.1371/journal.pone.0093466.

[NÖH 07] NÖH K., GRÖNKE K., LUO B. *et al.*, "Metabolic flux analysis at ultra short time scale: isotopically non-stationary 13C labeling experiments", *Molecular Systems Biology*, vol. 129, pp. 249–267, 2007. doi: 10.1016/j.jbiotec.2006.11.015.

[PAT 00] PATARINSKA T., DOCHAIN D., AGATHOS S.N. *et al.*, "Modelling of continuous microbial cultivation taking into account memory effects", *Bioprocess Engineering*, vol. 22, pp. 517–527, 2000.

[PER 60] PERRET J., "A new kinetic model of growing bacteria population", *Journal of General Microbiology*, vol. 2, pp. 589–617, 1960.

[PIG 15] PIGOU M., MORCHAIN J., "Investigating the interactions between physical and biological heterogeneities in bioreactors using compartment, population balance and metabolic models", *Chemical Engineering Science*, vol. 126, pp. 267–282, 2015. doi: 10.1016/j.ces.2014.11.035.

[PLA 78] PLASARI E., DAVID R., VILLERMAUX J., "Micromixing phenomena in continuous stirred reactors using a Michaelis–Menten reaction in the liquid phase", *Chemical Reaction Engineering – ACS Symposium Series*, pp. 125–139, 1978.

[RAM 71] RAMKRISHNA D., "Solution of population balance equations", *Chemical Engineering Science*, vol. 26, pp. 1134–1136, 1971.

[ROE 82] ROELS J.A., "Mathematical models and the design of biochemical reactors", *Journal of Chemical Technology and Biotechnology*, vol. 32, pp. 59–72, 1982. doi: 10.1002/jctb.5030320110.

[ROT 77] ROTENBERG M., "Selective synchrony of cells of differing cycle times", *Journal of Theoretical Biology*, vol. 66, pp. 389–398, 1977. doi: 10.1016/0022-5193(77)90179-5.

[ROT 83] ROTENBERG M., "Transport theory for growing cell populations", *Journal of Theoretical Biology*, vol. 103, pp. 181–199, 1983. doi: 10.1016/0022-5193(83)90024-3.

[SCH 03] SCHMALZRIEDT S., JENNE M., MAUCH K. *et al.*, "Integration of physiology and fluid dynamics", *Advances in Biochemical Engineering/Biotechnology*, Springer Berlin, Heidelberg, pp. 19–68, 2003.

[SCH 06] SCHÜTZE J., HENGSTLER J., "Assessing aerated bioreactor performance using CFD", *12th European Conference on Mixing*, Bologna, Italy, pp. 439–446, 2006.

[SCO 68] SCOTT W.T., "Analytic studies of cloud droplets coalescence", *Journal of the Atmospheric Sciences*, vol. 25, 1968.

[SEN 94] SENN H., LENDENMANN U., SNOZZI M. *et al.*, "The growth of *Escherichia coli* in glucose-limited chemostat cultures: a re-examination of the kinetics", *Biochimica et Biophysica Acta (BBA) – General Subjects*, vol. 1201, p. 424, 1994.

[SHO 03] SHOEMAKER J., REEVES G.T., GUPTA S. *et al.*, "The dynamics of single-substrate continuous cultures: the role of transport enzymes", *Journal of Theoretical Biology*, vol. 222, p. 307, 2003.

[SIL 08] SILVESTON P.L., BUDMAN H., JERVIS E., "Forced modulation of biological processes: a review", *Chemical Engineering Science*, vol. 63, pp. 5089–5105, 2008.

[SRI 99] SRIENC F., "Cytometric data as the basis for rigorous models of cell population dynamics", *Journal of Biotechnology*, vol. 71, pp. 233–238, 1999.

[STA 10] STAMATAKIS M., "Cell population balance, ensemble and continuum modeling frameworks: conditional equivalence and hybrid approaches", *Chemical Engineering Science*, vol. 65, pp. 1008–1015, 2010.

[STO 69] STORER F.F., GAUDY A.F., "Computational analysis of transient response to quantitative shock loadings of heterogeneous populations in continuous culture", *Environmental Science & Technology*, vol. 3, pp. 143–149, 1969.

[STR 91] STRAIGHT J.V., RAMKRISHNA D., "Complex growth dynamics in batch cultures: experiments and cybernetic models", *Biotechnology and Bioengineering*, vol. 37, pp. 895–909, 1991.

[SUB 70] SUBRAMANIAN G., RAMKRISHNA D., FREDRICKSON A. *et al.*, "On the mass distribution model for microbial cell populations", *Bulletin of Mathematical Biology*, vol. 32, pp. 521–537, 1970.

[SUN 12] SUNYA S., DELVIGNE F., URIBELARREA J.-L. *et al.*, "Comparison of the transient responses of *Escherichia coli* to a glucose pulse of various intensities", *Applied Microbiology and Biotechnology*, vol. 95, pp. 1021–1034, 2012.

[SWE 88] SWEERE A.P.J., GIESSELBACH J., BARENDSE R. *et al.*, "Modelling the dynamic behavior of *Saccharomyces cerevisiae* and its application in control experiments", *Applied Microbiology and Biotechnology*, vol. 28, 116–127, 1988. doi: 10.1007/BF00694298.

[SWE 89] SWEERE A.P.J., DALEN J.P., KISHONI E. *et al.*, "Theoretical analysis of the baker's yeast production: an experimental verification at a laboratory scale", *Bioprocess and Biosystems Engineering*, vol. 3, pp. 11–17, 1989.

[TAR 97] TARTAKOVSKY B., SHEINTUCH M., HILMER J.-M. *et al.*, "Modelling of *E. coli* fermentations: comparison of multicompartment and variable structure models", *Bioprocess Engineering*, vol. 16, pp. 323–329, 1997.

[TSA 71] TSAI B.I., FAN L.T., ERICKSON L.E. *et al.*, "The reversed two-environment model of micromixing and growth processes", *Journal of Applied Chemistry and Biotechnology*, vol. 21, pp. 307–312, 1971. doi: 10.1002/jctb.5020211008.

[VAN 86] VAN DIJKEN J.P., SCHEFFERS A.W., "Redox balances in the metabolism of sugars by yeasts", *FEMS Microbiology Reviews*, vol. 32, pp. 199–224, 1986.

[VAR 98] VARNER J., RAMKRISHNA D., "Application of cybernetic models to metabolic engineering: investigation of storage pathways", *Biotechnology and Bioengineering*, vol. 58, pp. 282–291, 1998.

[VAR 99] VARNER J., RAMKRISHNA D., "Mathematical models of metabolic pathways", *Current Opinion in Biotechnology*, vol. 10, pp. 146–150, 1999.

[VIL 72] VILLERMAUX J., DEVILLON J., "Représentation de la coalescence et de la redispersion des domaines de ségrégation dans un fluide par un modèle d'interaction phénoménologique", *Proceedings of the 2nd International Symposium on Chemical Reaction Engineering*, pp. 1–13, 1972.

[VIL 95] VILLERMAUX J., *Génie de la réaction chimique*, 2nd ed., TecDoc Lavoisier, 1995.

[VON 42] VON BERTALANFFY L., *Theoretische Biologie*, Gebrfider Borntraeger, Berlin-Zehlendorf, 1942.

[VRÁ 99] VRÁBEL P., VAN DER LANS R.G.J.M., CUI Y.Q. *et al.*, "Compartment model approach: mixing in large scale aerated reactors with multiple impellers", *Chemical Engineering Research and Design*, vol. 77, pp. 291–302, 1999. doi: 10.1205/026387699526223.

[VRÁ 00] VRÁBEL P., VAN DER LANS R.G.J.M., LUYBEN K.C.A.M. *et al.*, "Mixing in large-scale vessels stirred with multiple radial or radial and axial up-pumping impellers: modelling and measurements", *Chemical Engineering Science*, vol. 55, pp. 5881–5896, 2000.

[VRÁ 01] VRÁBEL P., VAN DER LANS R.G.J.M., VAN DER SCHOT F.N. *et al.*, "CMA: integration of fluid dynamics and microbial kinetics in modelling of large-scale fermentations", *Chemical Engineering Journal*, vol. 84, pp. 463–474, 2001. doi: 10.1016/S1385-8947(00)00271-0.

[WAD 05] WADLEY R., DAWSON M.K., "LIF measurements of blending in static mixers in the turbulent and transitional flow regimes", *5th International Symposium on Mixing in Industrial Processes (ISMIP5 60)*, pp. 2469–2478, 2005. doi: 10.1016/j.ces.2004.11.021.

[WIE 02] WIECHERT W., "Modeling and simulation: tools for metabolic engineering", *Journal of Biotechnology*, vol. 94, pp. 37–63, 2002.

[XIA 09] XIA J.-Y., WANG Y.-H., ZHANG S.-L. *et al.*, "Fluid dynamics investigation of variant impeller combinations by simulation and fermentation experiment", *Biochemical Engineering Journal*, vol. 43, pp. 252–260, 2009.

[YAS 11] YASUDA K., "Algebraic and geometric understanding of cells: epigenetic inheritance of phenotypes between generations", in MÜLLER S., BLEY T. (eds), *High Resolution Microbial Single Cell Analytics, Advances in Biochemical Engineering/Biotechnology*, Springer, Berlin, Heidelberg, 2011.

[YE 85] YE S.J., Micromixing in Saccharomyces cerevisiae aerobic fermentation, Thesis, Washington University, Saint Louis, 1985.

[YOU 70] YOUNG T.B., BRULEY D.F., BUNGAY H.R., "A dynamic mathematical model of the chemostat", *Biotechnology and Bioengineering*, vol. 12, pp. 747–769, 1970.

[YOU 08] YOUNG J.D., HENNE K.L., MORGAN J.A. *et al.*, "Integrating cybernetic modeling with pathway analysis provides a dynamic, systems-level description of metabolic control", *Biotechnology and Bioengineering*, vol. 100, pp. 542–559, 2008.

[ZHA 09] ZHANG H., ZHANG K., FAN S., "CFD simulation coupled with population balance equations for aerated stirred bioreactors", *Engineering in Life Sciences*, vol. 9, pp. 421–430, 2009.

Index

A, C, D

adaptation, 85, 95–97, 100, 102, 103, 111, 122, 131, 133, 134, 136, 141–143
age, 145–149, 165, 171–173
assimilation, 85–90, 92–96, 104–108, 110–112, 119, 128, 133–138, 143
average, 8–11, 14–16, 18, 19, 23–26
competition, 54–57, 71, 83
coupling, 168, 179, 184, 186
density function, 148, 149, 151, 164, 166, 179, 180
distribution, 33–35, 38, 39, 42, 43, 46, 65, 66, 68–73, 77, 82–84
dynamic, 88, 95, 97, 98, 100, 102, 104, 108, 111, 113, 115, 119, 121, 123, 127, 129, 134, 135, 137

E, G, H

equilibrium, 85, 87–89, 97, 100–103, 113–115, 122, 126, 131, 133, 135, 136, 138, 139, 141–143
growth rate, 156–159, 161, 165, 166, 168, 169, 174, 179–183, 185, 186
heterogeneous, 4, 6, 12, 22, 24, 159, 178, 184
homogeneous, 4, 6, 9, 11, 12, 26

I, L, M

ideal reactor, 8, 9
length scale, 45
limitation, 54, 60, 73
macromixing, 37, 40, 41, 44, 47, 48, 57–59, 84
mass, 145, 147, 149–157, 159–174, 179, 180
mass balances, 6, 7
micromixing, 37, 40, 44–46, 58–60, 77, 78, 80–83
model, 85, 113–116, 118–123, 125–127, 129–134, 136–138, 141, 143, 144
moment, 150, 162, 164–166, 169
multiphase, 6, 19, 24, 26

O, P, R

oxygen, 3–5, 12–17
perfectly mixed, 2, 4, 9, 10, 12
population, 145–151, 154–156, 158, 159, 165, 166, 168, 171–175, 177–187
property, 150, 151, 154
reaction rate, 31–34, 52, 55, 60–64, 66–70, 72

T, U

transfer, 2–4, 6–11, 13, 14, 16–19, 21–27, 85–91, 94, 128–130, 134, 135, 137, 143

turbulence, 30, 45, 48, 53, 56, 59, 64, 65, 79
uptake, 96, 97, 138, 143

Printed in the United States
By Bookmasters